STATISTICS OF
KNOTS AND ENTANGLED
RANDOM WALKS

STATISTICS OF KNOTS AND ENTANGLED RANDOM WALKS

S. K. Nechaev

Landau Insitute for Theoretical Physics
Russia
Institut de Physique Nucléaire
France

World Scientific
Singapore • New Jersey • London • Hong Kong

Published by

World Scientific Publishing Co. Pte. Ltd.

P O Box 128, Farrer Road, Singapore 912805

USA office: Suite 1B, 1060 Main Street, River Edge, NJ 07661

UK office: 57 Shelton Street, Covent Garden, London WC2H 9HE

British Library Cataloguing-in-Publication Data
A catalogue record for this book is available from the British Library.

STATISTICS OF KNOTS AND ENTANGLED RANDOM WALKS

ISBN 981-02-2519-9

This book is printed on acid-free paper.

Printed in Singapore by Uto-Print

PREFACE

It wouldn't be an exaggeration to say that contemporary physical science is becoming more and more mathematical. This fact is too strongly manifested to be completely ignored. One may argue about the origins of this phenomenon, but as to the question of how to account for this tendency, till now no sound hypothesis has been formed. Hence I would permit myself to bring forward three possible conjectures:

(a) There are hardly discovered any newly physical problem which would be beyond the well established methods of the modern theoretical physics;

(b) Modern mathematical physics is a fascinating field which generates a great deal of interest, especially in some of its rather artificial branches, what ultimately leads to invention of appropriate new languages;

(c) Nowadays real physical problems seem to be less numerous than mathematical methods of their investigation.

Certainly, I do not pretend to have arrived at the absolute truth and many of my colleagues—physicists—will not agree with this point of view, but I guess none of them will deny the fact of mentioned "mathematisation".

The penetration of new mathematical ideas in physics has sometimes rather paradoxical character. It is not a secret that difference in means (in languages) and goals of physicists and mathematicians leads to mutual misunderstanding, making the very subject of investigation obscure. What is true for general is certainly true for particular. To clarify the point, let us turn to statistics of entangled uncrossible random walks—the well-known subject of statistical physics of polymers. Actually, since 1970s, after Conway's works, when the first algebraic topological invariants—Alexander polynomials—became very popular in mathematical literature, physicists working in statistical topology have acquired a much more powerful topological invariant than the simple Gauss linking number, but nevertheless continue using the last one until recently making references to its imperfectness.

One of the reasons of such inertia consists in the fact that new mathematical ideas are often formulated as "theorems of existence" and it takes much time to retranslate them into physically acceptable form. Thus, shortly speaking, my purpose resembles more-or-less the interpreter's goal: I intend to use some recent advances in algebraic topology and theory of random walks on noncommutative groups for reconsidering the old problem— evaluating of the entropy of randomly generated knots and entangled random walks in a given homotopic state. I have tried to write a real physical book, replacing when it is possible the rigorous statements by some physically justified conjectures and applying them for investigation of thermodynamic properties of the given statistical systems. I should stress that despite a number of theorems in the text, this book offers far from standard ways of presentation of topological problems. Some of the ideas expressed below could be strongly criticized by topologists, but I will highly appreciate the response from mathematicians looking for physical applications of their topological ideas.

The book is devoted to analysis of probabilistic problems in topology. In order to establish the correlations between different statistical and topological problems, the following ones shall be discussed:

– Calculation of the probability for a long random walk to form randomly a knot with specific topological invariant. We consider the problem using the Kauffman algebraic invariants and show the connection with the thermodynamic properties of 2D Potts model with "quenched" and "annealed" disorder in interaction constants.

– Investigation of the limit behavior of random walks on the noncommutative groups related to the knot theory. Namely, we show the connection between the limit distribution for the Lyapunov exponent of products of noncommutative random matrices—generators of "braid group"—and the asymptotics of powers ("knot complexity") of algebraic knot invariants. The established relation is applied for calculation of the knot entropy. In particular, it is shown that the "knot complexity" corresponds to the well known topological invariant, "primitive path", repeatedly used in statistics of entangled polymer chains.

– Consideration of the random walks on multiconnected manifolds using conformal methods and construction of the nonabelian topological invariants. We show that many nontrivial properties of limit behavior of random walks with topological constraints can be explained in context of random walks on hyperbolic groups.

– Usage of the limit behavior of entangled random paths established above for investigation of: a) thermodynamic properties of uncrossible random walks in regular arrays of topological obstacles; b) statistical properties of so-called "crumpled globule" (trivial ring without self-intersections in strongly contracted state); c) stress-strain dependence in entangled polymer network and d) ordering phase transition in bunch of mutually entangled "directed polymers".

The body of the book is based on original works published during the last decade in different physical journals. Besides I express here some concepts elaborated through exchange of ideas with my colleagues.

I am very pleased to mention the names of physicists and mathematicians who should be undoubtedly regarded as direct co-authors of this book. These are: A. Grosberg, A. Khokhlov, V. Rostiashvili, A. Semenov, E. Shakhnovich, Ya. Sinai, A. Vershik. At the same time I should like to point out that the book is not a collection of known facts united by one subject. Some problems discussed here are well established, others, being in course of investigation, still are expressed as conjectures... Anyway, I hope that the work reflects, at least partially, the modern state of affairs in statistics of entangled chain-like objects.

The book consists of 4 Chapters: "Knot Diagrams as Disordered Spin Systems"; "Random Walks on Local Noncommutative Groups", "Conformal Methods in Statistics of Entangled Random Walks", "Physical Applications" and 2 Appendixes, which I found reasonable to represent as abridged versions of the already published papers: "Limit Theorem for Conditional Distribution of Products of Independent Unimodular 2×2 Matrices" (by S.K. Nechaev and Ya.G. Sinai) and "Polymer Chain in Random Array of Topological Obstacles" (by V.G. Rostiashvili, S.K. Nechaev and T. Vilgis). The connection between all these problems is shown in Table 1.

The original idea to publish some lecture notes about statistical topology belongs to Vik. Dotsenko who suggested that I arrange a sort of review on the subject, whereas the official offer to write a book came from World

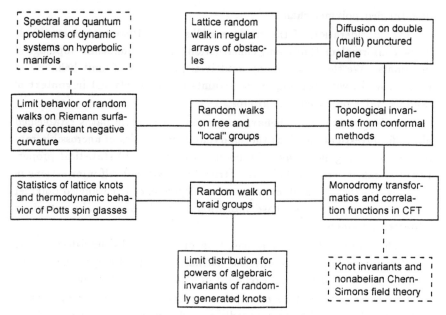

Table 1. Links between topologically-probabilistic problems. Solid boxes – problems, discussed in the book; dashed boxes – problems not included in the consideration.

Scientific Publishing Co. in 1994.

I express my sincere thanks to all mathematicians and physicists (in addition to those already mentioned) whose valuable remarks and suggestions helped me greatly in my work: E. Bogomolny, A. Comtet, J. Desbois, R. Dobrushin, Vik. Dotsenko, Vl. Dotsenko, B. Duplantier, I. Erukhimovich, M. Frank-Kamenetskii, Y. Georgelin, B. Hasslacher, F. Karpelevich, E. Kats, K. Khanin, A. Kitaev, D. Kuznetsov, I. Luck'yanchuk, V. Malyshev, A. Mazel, M. Monastyrsky, C. Monthus, S. Novikov, S. Obukhov, G. Oshanin, S. Ouvry, D. Parshin, Y. Rabin, N. Reshetikhin, M. Rubinstein, F. Ternovskii, A. Vologodskii, M. Yor. I am very grateful to V.F.R. Jones and L.H. Kauffman for some comments made after reading of the manuscript. Ms. A. Karpukhina rendered invaluable assistance in the text translation.

The major part of the book was written in Spring–Summer 1995 during my stay at Institut de Physique Nucléaire (Orsay, France) as a participant of a joint scientific program ENS (Paris)—Landau Institute (Moscow). Special mention should be made of the assistance given by M.

Mezard (ENS), V. Mineev (Landau Institute) and D. Vautherin (IPN) without whom the whole undertaking could hardly be accomplished. Finally I cannot end the acknowledgements without thanking the people at Landau Institute and IPN for spiritual support.

Sergei Nechaev
May 1996
Paris

CONTENTS

CHAPTER 2. RANDOM WALKS ON LOCAL NONCOMMUTATIVE GROUPS

CHAPTER 3. CONFORMAL METHODS IN STATISTICS OF ENTANGLED RANDOM WALKS

CHAPTER 4. PHYSICAL APPLICATIONS

CHAPTER 1

KNOT DIAGRAMS AS DISORDERED
SPIN SYSTEMS

1.1. Introduction: Statistical Problems in Topology

The interdependence of such branches of modern theoretical and mathematical physics as theory of integrable systems, algebraic topology and conformal field theory has proved to be a powerful catalyst of development of the new direction in topology, namely, of analytical topological invariants construction by means of exactly solvable statistical models.

Today it is widely believed that the following three cornerstone findings have brought the fresh stream in topology:

- The deep relation between the Temperley-Lieb algebra and the Hecke algebra representation of the braid group has been found. This fact resulted in the remarkable geometrical analogy between the Yang-Baxter equations, appearing as necessary condition of the transfer matrix commutativity in the theory of integrable systems on the one hand, and one of Reidemeister moves, used in the knot invariant construction on the other hand.

- It has been discovered that the partition function of the Wilson loop with the Chern-Simons action in the topological field theory coincides with the representation of the known nonabelian algebraic knot invariants written in terms of the time-ordered path integral.

1

- The need for new solutions of the Yang-Baxter equations has given rise to the theory of quantum groups. Later on the related set of problems was separated in the independent branch of mathematical physics.

Of course the above mentioned findings do not exhaust the list of all brilliant achievements in that field during the last decade, but apparently these new accomplishments have used profound "ideological" changes in the topological science: now we can hardly consider topology as an idependent branch of pure mathematics where each small step forward takes so much effort that it seems incidental.

Thus in the middle of the 80s the "quantum group" gin was released. It linked by common mathematical formalism classical problems in topology, statistical physics and field theory. A new look at the old problems and the beauty of the formulated ideas made an impression on physicists and mathematicians. As a result, in a few last years the number of works devoted to the search of the new applications of the quantum group apparatus is growing exponentially going beyond the framework of original domains. As an example of persistent penetrating of the quantum group ideas in physics we can name the works [*] on anyon superconductivity [1], intensively discussing problems on "quantum random walks" [2], the investigation of spectral properties of "quantum deformations" of harmonic oscillators [3] and so on.

The time will show whether such "quantum group expansion" is physically justified or it merely does tribute to today's fasion. However it is clear that physics has acquired new convenient language allowing to construct new "nonabeian objects" and to work with them.

Usually when starting the new paper (and especially a book), the author is asking himself the question: which place does his work take in the entire mass of investigations on the same subject. Frankly speaking, the author's own answer does not coincide not infrequently with that of the readers'... Hence before passing to the subject of the present book let us introduce the "coordinate system" for easier orientation in the related literature. Among the vast amount of works devoted to different aspects of the theory of integrable systems, its topological applications connected to the construction of knot and link invariants and their representation in

---

[*]Only the basis references are included in the lists.

terms of partition functions of some known 2D-models deserve our special attention.

There exist several reviews [4,5,6] and books [7,9] on the subject and our aim by no means consists in re-interpretation or compilation of their contents. In the present book we make the first attempt of consecutive account of recently solved *probabilistic* problems in topology as well as attract attention to some interesting, still unsolved, questions lying on the border of topology and the probability theory. Of course we employ the knowledges acquired in the algebraic topology utilizing the construction of new topological invariants done by V.F.R. Jones [4] and L.H. Kauffman [8].

Besides the traditional fundamental topological issues concerning the construction of new topological invariants, investigation of homotopic classes and fibre bundles we mark a set of ajoint but much less studied problems. First of all, we mean the problem of so-called "knot entropy" calculation. Most generally it can be formulated as follows. Take the lattice $\mathbf{Z}^3$ embeded in the space $\mathbf{R}^3$. Let Ω_N be the ensemble of all possible closed nonselfintersecting N-step loops with one common fixed point on $\mathbf{Z}^3$; by ω we denote the particular trajectory configuration. The question is: what is the probability $\mathcal{P}_N$ of the fact that the trajectory $\omega \in \Omega_N$ belongs to some specific homotopic class. Formally this quantity can be represented in the following way

$$
\mathcal{P}_N\{\mathrm{Inv}\} = \frac{1}{\Omega_N} \sum_{\{\omega\}} \Delta\left[\mathrm{Inv}\{\omega\} - \mathrm{Inv}\right] \equiv
$$

$$
\frac{1}{\Omega_N} \sum_{\{\mathbf{r}_1,\ldots,\mathbf{r}_N\}} \Delta\left[\mathrm{Inv}\{\mathbf{r}_1,\ldots,\mathbf{r}_N\} - \mathrm{Inv}\right] \left(1 - \Delta\left[\mathbf{r}_i - \mathbf{r}_j\right]\right) \Delta\left[\mathbf{r}_N\right]
$$

(1.1)

where $\mathrm{Inv}\{\omega\}$ is the functional representation of the knot invariant corresponding to the trajectory with the bond coordinates $\{\mathbf{r}_1 \ldots, \mathbf{r}_N\}$; Inv is the topological invariant characterizing the knot of specific homotopic type and $\Delta(x)$ is the Kronecker function: $\Delta(x = 0) = 1$ and $\Delta(x \neq 0) = 0$. The first Δ-function in Eq.(1.1) cuts the set of trajectories with the fixed topological invariant while the second and the third Δ-functions ensure the N-step trajectory to be nonselfintersecting and to form a closed loop respectively.

The distribution function $\mathcal{P}_N\{\mathrm{Inv}\}$ satisfies the normalization condition

$$\sum_{\text{all homotopic classes}} P_N\{\text{Inv}\} = 1 \qquad (1.2)$$

Definition 1 *The entropy $S_N\{\text{Inv}\}$ of the given homotopic state of the knot represented by N-step closed loop on $\mathbf{Z}^3$ reads*

$$S_N\{\text{Inv}\} = \ln\left[\Omega_N \mathcal{P}_N\{\text{Inv}\}\right] \qquad (1.3)$$

The problem concerning the knot entropy determination has been discussed time and again by the leading physicists. However the number of new analytic results in that field was insufficient till the beginning of the 80s: in about 90 percents of published materials their authors used the Gauss linking number or some of its abelian modifications for classification of a topological state of knots and links while the disadvantages of this approach were explained in the rest 10 percent of the works. We do not include in this list the celebrated investigations of A.V. Vologodskii *et al* [10] devoted to the first fruitful usage of the nonabelian Alexander algebraic invariants for the computer simulations in the statistical biophysics. We discuss physical applications of these topological problems at length in Chapter 4.

Topological ideas are very easy to understand from the geometrical point of view but they are hard to formalize because of the non-local character of topological constraints. Besides, the main difficulty in attempts to calculate analytically the knot entropy is due to the absence of convenient analytic representation of the complete topological invariant. Thus, to succeed, at least partially, in the knot entropy computation we simplify the general problem (see Definition 1) replacing it by the problem of calculation of the distribution function for the knots *with defined topological invariants*. That problem differs from the original one because none of the known topological invariants (Gauss linking number, Alexander, Jones, HOMFLY) are complete. The only exception is Vassiliev invariants [11], which are beyond the scope of the present book. Strictly speaking we are unable to estimate exactly the correctness of such replacement of the homotopic class by the mentioned topological invariants. Thus under the definition of the topological state of the knot or entanglement we simply understand the determination of the corresponding topological invariant.

The problems where ω (see Eq.(1.1)) is the set of realizations of the random walk, i.e. the Markov chain are of special interest. In that case the probability to find a closed N-step random walk in $\mathbf{R}^3$ in some prescribed topological state can be presented in the following way

$$P_N\{\text{Inv}\} = \int \ldots \int \prod_{j=1}^{N} d\mathbf{r}_j \prod_{j=1}^{N-1} g\left(\mathbf{r}_{j+1} - \mathbf{r}_j\right) \delta\left[\text{Inv}\{\mathbf{r}_1 \ldots, \mathbf{r}_N\} - \text{Inv}\right] \delta\left[\mathbf{r}_N\right]$$

$$(1.4)$$

where $g\left(\mathbf{r}_{j+1} - \mathbf{r}_j\right)$ is the probability to find $j + 1$th step of the trajectory in the point $\mathbf{r}_{j+1}$ if jth step is in $\mathbf{r}_j$. In the limit $a \to 0$ and $N \to \infty$ ($Na = L = \text{const}$) in three-dimensional space we have the following expression for $g\left(\mathbf{r}_{j+1} - \mathbf{r}_j\right)$

$$
\begin{aligned}
g\left(\mathbf{r}_{j+1} - \mathbf{r}_j\right) &= \frac{1}{4\pi a^2} \delta\left(\left|\mathbf{r}_{j+1} - \mathbf{r}_j\right| - a\right) \\
&\simeq \left(\frac{3}{2\pi a^2}\right)^{3/2} \exp\left(-\frac{3(\mathbf{r}_{j+1} - \mathbf{r}_j)^2}{2a^2}\right)
\end{aligned}
$$

$$(1.5)$$

Introducing the "time", s, along the trajectory we rewrite the distribution function $P_N\{\text{Inv}\}$ (Eq.(1.4)) in the path inegral form with the Wiener measure density

$$P_N\{\text{Inv}\} = \frac{1}{Z} \int \ldots \int \mathcal{D}\{\mathbf{r}\} \exp\left\{-\frac{3}{2a^2} \int_0^L \left(\frac{d\mathbf{r}(s)}{ds}\right)^2 ds\right\}$$

$$\times \delta[\text{Inv}\{\mathbf{r}(s)\} - \text{Inv}]$$

$$(1.6)$$

where the following sequence of limits in the exponent is taken

$$
\begin{aligned}
\lim_{\substack{N \to \infty \\ a \to 0}} \sum_{j=1}^{N-1} \frac{3(\mathbf{r}_{j+1} - \mathbf{r}_j)^2}{2a^2} &= \lim_{\substack{N \to \infty \\ a \to 0}} \frac{3}{2a} \sum_{s=a}^{Na} \left(\frac{\mathbf{r}(s+a) - \mathbf{r}(s)}{a}\right)^2 a \\
&= \frac{3}{2a} \int_0^L \left(\frac{d\mathbf{r}(s)}{ds}\right)^2 ds
\end{aligned}
$$

$$(1.7)$$

and the normalization condition is as follows

$$Z = \sum_{\substack{\text{all different} \\ \text{knot invariants}}} P_N\{\text{Inv}\}$$

The form of Eq.(1.6) up to the Wick turn and the constants coincides with the scattering amplitude α of a free quantum particle in the multiconnected phase space. Actually, for the amplitude α we have

$$\alpha \sim \sum_{\substack{\text{all paths from given} \\ \text{topological class}}} \exp\left\{\frac{i}{h}\int \dot{\mathbf{r}}^2(s)ds\right\} \tag{1.8}$$

If phase trajectories can be mutually transformed by means of continuous deformations, then the summation in Eq.(1.8) should be extended to all available paths in the system, but if the phase space consists of different topological domains, then the summation in Eq.(1.8) refers to the paths from the exclusively defined class and the "knot entropy" problem arises.

1.2. Review of Abelian Problems in Statistics of Entangled Random Walks and Incompleteness of Gauss Invariant

As far back as 1967 S.F. Edwards had discovered the basis of the statistical theory of entanglements in physical systems. In [12] he proposed the way of exact calculation of the partition function of selfintersecting random walk topologically interacting with the infinitely long uncrossible string (in 3D case) or obstacle (in 2D-case). That problem had been considered in mathematical literature even earlier—see the paper [13] for instance—but S.F. Edwards was apparently the first to recognize the deep analogy between abelian topological problems in statistical mechanics of the Markov chains and quantum-mechanical problems (like Bohm-Aharonov) of the particles in the magnetic fields. The review of classical results is given in [15], whereas some modern advantages are discussed in [14].

The 2D version of the Edwards' model is formulated as follows. Take a plane with an excluded origin, producing the topological constraint for the random walk of length L with the initial and final points $\mathbf{r}_0$ and $\mathbf{r}_L$ respectively. Let trajectory make n turns around the origin (fig.1.1). The question is in calculating the distribution function $\mathcal{P}_n(\mathbf{r}_0, \mathbf{r}_L, L)$.

For the said model the topological state of the path C is fully characterized by number of turns of the path around the origin. The corresponding abelian topological invariant is known as Gauss linking number and when represented in the countor integral form, reads

$$\text{Inv}\{\mathbf{r}(s)\} \equiv G\{C\} = \int_C \frac{ydx - xdy}{x^2 + y^2} = \int_C \mathbf{A}(\mathbf{r})d\mathbf{r} \equiv 2\pi n + \vartheta \tag{1.9}$$

where

$$\mathbf{A}(\mathbf{r}) = \boldsymbol{\xi} \times \frac{\mathbf{r}}{r^2}; \qquad \boldsymbol{\xi} = (0, 0, 1) \tag{1.10}$$

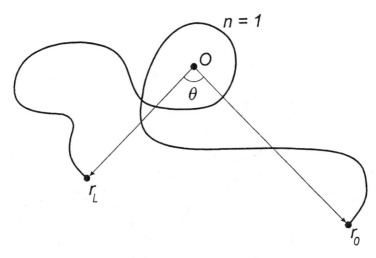

Fig. 1.1. Random walk on the plane near the single obstacle.

and ϑ is the angle distance between ends of the random walk.

Substituting Eq.(1.9) into Eq.(1.6) and using the Fourier transform of the δ-function, we arrive at

$$\mathcal{P}_n(\mathbf{r}_0, \mathbf{r}_L, L) = \frac{1}{\pi La} \exp\left(\frac{r_0^2 + r_L^2}{La}\right) \int_{-\infty}^{\infty} I_{|\lambda|}\left(\frac{2r_0 r_L}{La}\right) e^{i\lambda(2\pi n + \vartheta)} d\lambda$$

(1.11)

which reproduces the well known old result [12] (some very important generalizations one can find in [14]).

Physically significant quantity obtained on the basis of Eq.(1.11) is the entropic force

$$f_n(\rho) = -\frac{\partial}{\partial \rho} \ln \mathcal{P}_n(\rho, L)$$

(1.12)

which acts on the closed chain ($\mathbf{r}_0 = \mathbf{r}_L = \rho$, $\vartheta = 0$) when the distance between the obstacle and a certain point of the trajectory changes. Apparently the topological constraint leads to the strong attraction of the path to the obstacle for any $n \neq 0$ and to the weak repulsion for $n = 0$.

Another exactly solvable 2D-problem closely related to the one under discussion deals with the calculation of the partition function of a random walk with given algebraic area. The problem concerns the determination of the distribution function $\mathcal{P}_S(\mathbf{r}_0, \mathbf{r}_L, L)$ for the random walk with the fixed ends and specific algebraic area S (see fig.1.2).

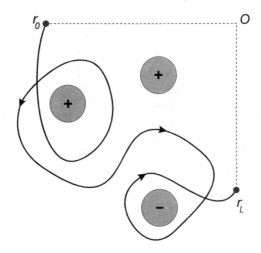

Fig. 1.2. Random walk with fixed algebraic area.

As a possible solution of that problem, D.S. Khandekar and F.W. Wiegel [16] again represented the distribution function in terms of the path integral Eq.(1.6) with the replacement

$$\delta[\mathrm{Inv}\{\mathbf{r}(s)\} - \mathrm{Inv}] \rightarrow \delta[S\{\mathbf{r}(s)\} - S] \qquad (1.13)$$

where the area is written in the Landau gauge:

$$S\{\mathbf{r}(s)\} = \frac{1}{2}\int_C y\,dx - x\,dy = \frac{1}{2}\int_C \tilde{\mathbf{A}}\{\mathbf{r}\}\dot{\mathbf{r}}\,ds; \qquad \tilde{\mathbf{A}} = \boldsymbol{\xi} \times \mathbf{r} \qquad (1.14)$$

(compare to Eqs.(1.9)-(1.10)).

The final expression for the distribution function reads ([15])

$$\mathcal{P}_S(\mathbf{r}_0, \mathbf{r}_L, L) = \frac{1}{2\pi}\int_{-\infty}^{\infty} dg\; e^{iqS}\, \mathcal{P}_q(\mathbf{r}_0, \mathbf{r}_L, L) \qquad (1.15)$$

where

$$\mathcal{P}_q(\mathbf{r}_0, \mathbf{r}_L, L) = \frac{\lambda}{4\pi\sin\frac{La\lambda}{4}}$$
$$\times \exp\left\{\frac{\lambda}{2}(x_0 y_L - y_0 x_L) - \frac{\lambda}{4}\left((x_L - x_0)^2 + (y_L - y_0)^2\right)\cot\frac{La\lambda}{4}\right\}$$
$$(1.16)$$

and $\lambda = -iq$.

For closed trajectories Eqs.(1.15)-(1.16) can be simplified essentially, giving

$$\mathcal{P}_S^{cl}(N) = 2La\cosh^2\left(\frac{2\pi S}{La}\right) \qquad (1.17)$$

Different aspects of this problem have been extensively studied in [18].

There is no principal difference between the problems of random walk statistics in the presence of a single topological obstacle or with a fixed algebraic area—both of them have the "abelian" nature. Nevertheless we would like to concentrate on the last problem because of its deep connection with the famous Harper-Hofstadter model dealing with spectral properties of the 2D electron hopping on the discrete lattice in the constant magnetic field [17]. Actually, rewrite Eq.(1.4) with the substitution Eq.(1.13) in form of recursion relation in the number of steps, N:

$$\mathcal{P}_q(\mathbf{r}_{N+1}, N+1) = \int d\mathbf{r}_N g\left(\mathbf{r}_{N+1} - \mathbf{r}_N\right)\exp\left(\frac{iq}{2}\xi(\mathbf{r}_N \times \mathbf{r}_{N+1})\right)$$
$$\times \mathcal{P}_q(\mathbf{r}_N, N) \qquad (1.18)$$

For the discrete random walk on $\mathbb{Z}^2$ we use the identity

$$\int d\mathbf{r}_N g\left(\mathbf{r}_{N+1} - \mathbf{r}_N\right)(\ldots) \rightarrow \sum_{\{\mathbf{r}_N\}} w\left(\mathbf{r}_{N+1} - \mathbf{r}_N\right)(\ldots) \qquad (1.19)$$

where $w\left(\mathbf{r}_{N+1} - \mathbf{r}_N\right)$ is the martix of the local jumps on the square lattice; w is supposed to be symmetric:

$$w = \begin{cases} \frac{1}{4} & \text{for } (x,y) \rightarrow (x, y \pm 1) \text{ and } (x,y) \rightarrow (x \pm 1, y) \\ 0 & \text{otherwise} \end{cases} \qquad (1.20)$$

Finally, we get in the Landau gauge:

$$\frac{4}{\varepsilon}W(x, y, q, \varepsilon) = e^{\frac{1}{2}iqx}W(x, y-1, q) + e^{-\frac{1}{2}iqx}W(x, y+1, q) +$$
$$e^{\frac{1}{2}iqy}W(x-1, y, q) + e^{-\frac{1}{2}iqy}W(x+1, y, q) \qquad (1.21)$$

where $W(x, y, q, \varepsilon)$ is the generating function defined via relation

$$W(x, y, q, \varepsilon) = \sum_{N=0}^{\infty} \varepsilon^N \mathcal{P}_S(\mathbf{r}_N, N)$$

and q plays a role of the magnetic flux through the contour bounded by the random walk on the lattice.

There is one point which is still out of our complete understanding. On the one hand the continuous version of the described problem has very clear abelian background due to the use of commutative "invariants" like algebraic area Eq.(1.14). On the other hand it has been recently discovered ([19]) that so-called Harper equation, i.e. Eq.(1.21) written in the gauge $S\{\mathbf{r}\} = \int_C y dx$, exhibits the hidden quantum group symmetry related to the so-called C^*–algebra ([20]) which is strongly nonabelian. Usually in statistical physics we expect that the continuous limit (when lattice spacing tends to zero with corresponding rescaling of parameters of the model) of any discrete problem does not change the observed physical picture, at least qualitatively. But for the considered model the spectral properties of the problem are extremely sensitive to the actual physical scale of the system and depend strongly on the lattice geometry.

INCOMPLETENESS OF GAUSS INVARIANT. The generalization of the above stated problems concerns, for instance, the calculation of the partition function for the random walk entangled with $k > 1$ obstacles on the plane located in the points $\{\mathbf{r}_1, \ldots, \mathbf{r}_k\}$. At first sight, approach based on usage of Gauss linking number as topological invariant, might allow us to solve such problem easily. Let us replace the vector potential $\mathbf{A}(\mathbf{r})$ in Eq.(1.9) by the following one

$$\mathbf{A}(\mathbf{r}_1, \ldots, \mathbf{r}_k) = \xi \times \sum_{j=1}^{k} \frac{\mathbf{r} - \mathbf{r}_j}{|\mathbf{r} - \mathbf{r}_j|^2} \qquad (1.22)$$

The topological invariant in this case will be the algebraic sum of turns around obstacles, which seems to be a natural generalization of the Gauss linking number to the case of many-obstacle entanglements.

However, the following problem is bound to arise: for the system with two or more obstacles it is possible to imagine closed trajectories entangled with a few obstacles together but not entangled with every one. In the fig.1.3 the so-called "Pochammer contour" is shown. Its topological state with respect to the obstacles cannot be described using any abelian version of the Gauss-like invariants.

To clarify the point we can apply to the concept of the homotopy group [21]. Consider the topological space $\mathcal{R} = \mathbf{R}^2 - \{\mathbf{r}_1, \mathbf{r}_2\}$ where $\{\mathbf{r}_1, \mathbf{r}_2\}$ are the coordinates of the removed points (obstacles) and choose an arbi-

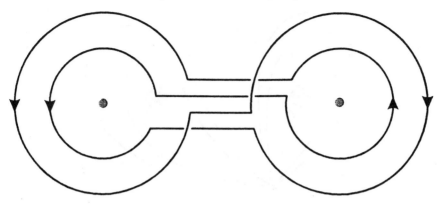

Fig. 1.3. Pochammer contour entangled with two obstacles together but not entangled with every one.

trary reference point $\mathbf{r}_0$. Consider the ensemble of all directed trajectories starting and finishing in the point $\mathbf{r}_0$. Take the *basis loops* $\gamma_1(s)$ and $\gamma_2(s)$ $(0 < s < L)$ representing the right-clock turns around the points $\mathbf{r}_1$ and $\mathbf{r}_2$ respectively. The same trajectories passed in the counter-clock direction are denoted by $\gamma_1^{-1}(s)$ and $\gamma_2^{-1}(s)$ — see fig.1.4.

The *multiplication* of the paths is their composition: for instance, $\gamma_1\gamma_2 = \gamma_1 \circ \gamma_2$. The unit (trivial) path is the composition of an arbitrary loop with its inverse:

$$e = \gamma_i\gamma_i^{-1} = \gamma_i^{-1}\gamma_i \qquad\qquad i = \{1,2\}. \qquad\qquad (1.23)$$

The loops $\gamma_i(s)$ and $\tilde{\gamma}_i(s)$ are called equivalent if one can be transformed into another by means of monotonic change of variables $s = s(\tilde{s})$. The homotopic classes of directed trajectories form the group with respect to the paths multiplplication; the unity is the homotopic class of the trivial paths. This group is known as the *homotopy group* $\pi_1(\mathcal{R}, \mathbf{r}_0)$.

Any closed path on $\mathcal{R}$ can be represented by the "word" consisting of set of letters $\{\gamma_1, \gamma_2, \gamma_1^{-1}, \gamma_2^{-1}\}$. Taking into account Eq.(1.23), we can reduce each word to the minimal irreducible representation. For example, the word $W = \gamma_1\gamma_2^{-1}\gamma_1\gamma_1\gamma_1^{-1}\gamma_2^{-1}\gamma_2\gamma_1^{-1}\gamma_2^{-1}$ can be transformed to the irreducible form: $W = \gamma_1\gamma_2^{-1}\gamma_2^{-1}$. It is easy to understand that the word $W \equiv e$ represents only the unentangled contours. The entanglement in fig.1.3 corresponds to the irreducible word $W = \gamma_1^{-1}\gamma_2\gamma_1\gamma_2^{-1} \equiv 1$. The non-abelian character of the topological constraints is reflected in the fact

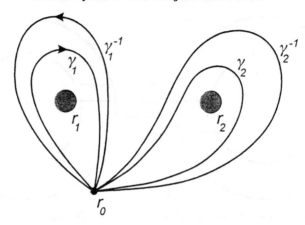

Fig. 1.4. Basis loops as the generators of the fundamental group π_1.

that different entanglements do not commute: $\gamma_1\gamma_2 \neq \gamma_2\gamma_1$. At the same time, the total algebraic number of turns (Gauss linking number) for the path in fig.1.3 is equal to zero, i.e. it belongs to the trivial *class of homology*. Speaking more formally, the mentioned example is the direct consequence of the well known fact in topology: the classes of homology of knots (of entanglements) do not coincide in general with the corresponding homotopic classes. The first ones for the group π_1 can be distingushed by the Gauss invariant, while the problem of characterizing the homotopy class of a knot (entanglement) by an analytically defined invariant is one of the main problems in topology. The related questions are discussed at length in the Chapter 3.

The principal difficulty connected with application of the Gauss invariant is due to its incompleteness. Hence, exploiting the abelian invariants for adequate classification of topologically different states in the systems with multiple topological constraints is very problematic. Nevertheless this approach is repeatedly used for consideration of such matters as the high-elasticity of polymer networks with topological constraints [22]. Unfortunately these works disregard the correctness of the Gauss linking number application and the question about their validity remains to be open.

1.3. Nonabelian Algebraic Knot Invariants

The most obvious topological questions concerning the knotting probability during the random closure of the random walk cannot be answered using the Gauss invariant due to its weakness.

The break through in that field was made in 1975-1976 when the algebraic polynomials were used for the topological state identification of closed random walks generated by the Monte-Carlo method [10]. It has been recognized that the Alexander polynomials being much stronger invariants than the Gauss linking number, could serve as a convenient tool for the calculation of the thermodynamic properties of entangled random walks. That approach actually appeared to be very fruitful and the main part of our modern knowledge on knots and links statistics is obtained with the help of these works and their subsequent modifications.

In the present Chapter we develop the analytic approach in statistical theory of knots considering the basic problem—the probability to find a randomly generated knot in a specific topological state. We would like to reiterate that our investigation would be impossible without utilizing of algebraic knot invariants discovered recently. Below we reproduce briefly the construction of Jones invariants following the Kauffman approach in the general outline.

1.3.1. *Disordered Potts Model and Generalized Dichromatic Polynomials*

The graph expansion for the Potts model with the disorder in the interaction constants can be defined by means of slight modification of the well known construction of the ordinary Potts model [23,24]. Let us recall the necessary definitions.

Take an arbitrary graph $\mathcal{L}$ with N vertices. To each vertex of the given graph we attribute the "spin" variable σ_i ($i \in [1, N]$) which can take q states labelled as $1, 2, \ldots, q$ on the simplex. Suppose that the interaction between spins belonging to the connected neighboring graph vertices only contributes to the energy. Define the energy of the spin's interaction as follows

$$E_{kl} = J_{kl}\, \delta(\sigma_k, \sigma_l) = \begin{cases} J_{kl} & \sigma_k = \sigma_l,\ (\sigma_k, \sigma_l) - \text{neighbors} \\ 0 & \text{otherwise} \end{cases} \tag{1.24}$$

where J_{kl} is the interaction constant which varies for different graph edges and the equality $\sigma_k = \sigma_l$ means that the neighboring spins take equal values on the simplex.

The partition function of the Potts model now reads

$$Z_{potts} = \sum_{\{\sigma\}} \exp \left\{ \sum_{\{kl\}} \frac{J_{kl}}{T} \delta(\sigma_k, \sigma_l) \right\} \qquad (1.25)$$

where T is the temperature.

Expression Eq.(1.25) gives for $q = 2$ the well-known representation of the Ising model with the disordered interactions extensively studied in the theory of spin glasses [25]. (Later on we would like to fill in this old story by a new "topological" sense.)

To proceed with the graph expansion of the Potts model [24], rewrite the partition function (1.25) in the following way

$$Z_{potts} = \sum_{\{\sigma\}} \prod_{\{kl\}} [1 + v_{kl}\, \delta(\sigma_k, \sigma_l)] \qquad (1.26)$$

where

$$v_{kl} = \exp \left(\frac{J_{kl}}{T} \right) - 1 \qquad (1.27)$$

If the graph $\mathcal{L}$ has N edges then the product Eq.(1.26) contains N multipliers. Each multiplier in that product consists of two terms $\{1$ and $v_{kl}\, \delta(\sigma_k, \sigma_l)\}$. Hence the partition function Eq.(1.26) is decomposed in the sum of 2^N terms.

Each term in the sum is in one-to-one correspondence with some part of the graph $\mathcal{L}$. To make this correspondence clearer, it should be that an arbitrary term in the considered sum represents the product of N multipliers described above in ones from each graph edge. We accept the following convention:

i) If for some edge the multiplier is equal to 1, we remove the corresponding edge from the graph $\mathcal{L}$;

ii) If the multiplier is equal to $v_{kl}\, \delta(\sigma_k, \sigma_l)$ we keep the edge in its place.

After repeating the same procedure with all graph edges, we find the unique representation for all terms in the sum Eq.(1.26) by collecting the components (either connected or not) of the graph $\mathcal{L}$.

Take the typical graph G consisting of m edges and C connected components where the separated graph vertex is considered as one component. The presence of δ-functions ensures the spin's equivalence within one graph component. As a result after summation of all independent spins and of all possible graph decompositions we get the new expression for the partition function of the Potts system Eq.(1.25)

$$Z_{potts} = \sum_{\{G\}} q^C \prod_{\{kl\}}^{m} v_{kl} \qquad (1.28)$$

where the product runs over all edges in the fixed graph G.

It should be noted that the graph expansion Eq.(1.28) where $v_{kl} \equiv v$ for all $\{k,l\}$ coincides with the well known representation of the Potts system in terms of *dichromatic polynomial* (see, for instance, [23,24,27]).

Another comment concerns the number of spin states, q. As it can be seen, in the derivation presented above we did not account for the fact that q has to take positive integer values only. From this point of view the representation Eq.(1.28) has an advantage with respect to the standard representation Eq.(1.25) and can be considered an analytic continuation of the Potts system to the noninteger and even complex values of q. We show in the subsequent sections how the defined model is connected to the algebraic knot invariants.

1.3.2. *Reidemaister Moves and State Model for Construction of Algebraic Invariants*

Let K be a knot (or link) embedded in the 3D-space. First of all we project the knot (link) onto the plane and obtain the 2D-knot diagram in the so-called general position (denoted by K as well). It means that only the pair crossings can be in the points of paths intersections. Then for each crossing we define the passages, i.e. parts of the trajectory on the projection going "below" and "above" in accordance with its natural positions in the 3D-space.

For the knot plane projection with defined passages the following theorem is valid:

Theorem 1 (Reidemeister [28]) *Two knots embedded in* $\mathbf{R}^3$ *can be deformed continuously one into the other if and only if the diagram of one knot can be transformed into the diagram corresponding to another knot via the sequence of simple local moves of types* I, II *and* III *shown in* fig.1.5.

The work [28] provides us with the proof of this theorem. Two knots are called *regular isotopic* if they are isotopic with respect to two last Reidemeister moves (II and III); meanwhile, if they are isotopic with respect to

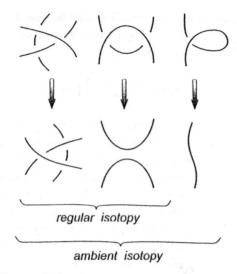

Fig. 1.5. Redemeister moves of types I, II and III.

all moves, they are called *ambient isotopic*. As it can be seen from fig.1.5, the Reidemeister move of type I leads to the cusp creation on the projection. At the same time it should be noted that all real 3D-knots (links) are of ambient isotopy.

Now, after the Reidemeister theorem has been formulated, it is possible to describe the construction of polynomial "bracket" invariant in the way proposed by L.H. Kauffman [8,9,29]. This invariant can be introduced as a certain partition function being the sum over the set of some formal ("ghost") degrees of freedom.

Let us consider the 2D-knot diagram with defined passages as a certain irregular lattice (graph). Crossings of path on the projection are the lattice vertices. Turn all these crossings to the standard positions where parts of the trajectories in each graph vertex are normal to each other and form the angles of $\pm\pi/4$ with the x-axis. It can be proven that the result does not depend on such standardization.

There are two types of vertices in our lattice—a) and b) which we label by the variable $b_i = \pm 1$ as it is shown below:

a) $\times$ $b_i = +1$ and   b) $\times$ $b_i = -1$

The next step in the construction of algebraic invariant is introduction of two possible ways of *vertex splittings*. Namely, we attribute to each way of graph splitting the following statistical weights:

- *A* to the horizontal splitting and *B* to the vertical one for the vertex of type a);

- *B* to the horizontal splitting and *A* to the vertical one for the vertex of type b).

The said can be schematically reproduced in the following picture:

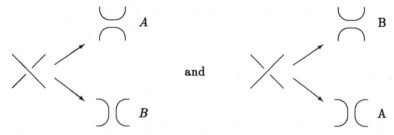

the constants *A* and *B* to be defined later.

For the knot diagram with N vertices there are 2^N different microstates, each of them representing the set of splittings of all N vertices. The entire microstate, S, corresponds to the knot (link) disintegration to the system of disjoint and non-selfintersecting circles. The number of such circles for the given microstate S we denote as $\mathcal{S}$.

Theorem 2 (Kauffman [9]**)** *Consider the partition function*

$$\langle K \rangle = \sum_{\{S\}} d^{\mathcal{S}-1} A^i B^j, \qquad (1.29)$$

where $\sum_{\{S\}}$ means summation over all possible 2^N graph splittings, i and $j = N - i$ being the numbers of vertices with weights A and B for the given realization of all splittings in the microstate S respectively.

The polynomial in A, B and d represented by the partition function Eq.(1.29) is the topological invariant of knots of regular isotopy if and only if the following relations among the weights A, B and d are fulfilled:

$$AB = 1$$
$$ABd + A^2 + B^2 = 0 \qquad (1.30)$$

Proof. Denote with $\langle \ldots \rangle$ the statistical weight of the knot or of its part. The $\langle K \rangle$-value equals the product of all weights of knot parts. Using the definition of vertex splittings, it is easy to test the following identities valid for unoriented knot diagrams

$$\left\langle \times \right\rangle = \left\langle \asymp \right\rangle A + \left\langle)(\right\rangle B \qquad (1.31)$$

$$\left\langle \times \right\rangle = \left\langle \asymp \right\rangle B + \left\langle)(\right\rangle A$$

completed by the "initial condition"

$$\left\langle K \bigcup O \right\rangle = d \left\langle K \right\rangle; \qquad K \text{ is not empty} \qquad (1.32)$$

where O denotes the separated trivial loop.

The *skein relations* Eq.(1.31) correspond to the above defined weights of horizontal and vertical splittings while the relation Eq.(1.32) defines the statistical weights of the composition of an arbitrary knot and a single trivial ring. These diagrammatic rules are well defined only for fixed "boundary condition" of the knot (i.e., for the fixed part of the knot outside the brackets). Suppose that by convention the polynomial of the trivial ring is equal to the unity;

$$\left\langle O \right\rangle = 1 \qquad (1.33)$$

Now it can be shown that under the appropriate choice of the relations between A, B and d, the partition function Eq.(1.29) represents the algebraic invariant of the knot. The proof is based on direct testing of the invariance of $\langle K \rangle$-value with respect to the Reidemeister moves of types II and III. For instance, for the Reidemeister move of type II we have:

$$\left\langle \smile\!\!\frown \right\rangle = \left\langle \asymp\!\circ \right\rangle ABd + \left\langle \bowtie \right\rangle B^2 +$$

$$\left\langle \bowtie \right\rangle A^2 + \left\langle \rangle\langle \right\rangle AB =$$

$$\left\langle \smile\!\!\frown \right\rangle (ABd + A^2 + B^2) + \left\langle \,)\,(\right\rangle AB$$

$$(1.34)$$

　　Therefore, the invariance with respect to the Reidemeister move of type II can be obtained immediately if we set the statistical weights in the last line of Eq.(1.34) as it is written in Eq.(1.30). Actually, the topological equivalence of two knot diagrams is restored with respect to the Reidemeisere move of type II only if the right- and left-hand sides of Eq.(1.34) are identical.

　　It can also be tested that the condition of obligatory invariance with respect to the Reidemeister move of type III does not violate the relations Eq.(1.30):

$$\left\langle X \right\rangle = \left\langle \asymp \right\rangle A + \left\langle \rangle\langle \right\rangle B =$$

$$\left\langle \asymp \right\rangle A + \left\langle \rangle\langle \right\rangle B = \left\langle X \right\rangle$$

$$(1.35)$$

Thus, the choice of statistical weights as in Eq.(1.30) is consistent with Eq.(1.35) and the theorem is proved □.

　　The relations Eq.(1.30) can be converted into the form

$$B = A^{-1}, \qquad d = -A^2 - A^{-2} \qquad (1.36)$$

which means that the Kauffman invariant Eq.(1.29) is the Laurent polynomial in A-value only.

Finally, Kauffman showed that for oriented knots (links) the invariant of ambient isotopy (i.e., the invariant with respect to all Reidemeister moves) is defined via relation:

$$f[K] = (-A)^{3Tw(K)} \langle K \rangle \qquad (1.37)$$

here $Tw(K)$ is the twisting of the knot (link), i.e. the sum of signs of all crossings defined by the convention:

a) $+1$ b) -1

(not to be confused with the definition of the variable b_i introduced above). Eq.(1.37) follows from the following chain of equalities

$$f\left[\begin{matrix} \end{matrix} \right] = \left\langle \begin{matrix} \end{matrix} \right\rangle B + \left\langle \begin{matrix} \end{matrix} \right\rangle dA$$

$$= \left\langle \begin{matrix} \end{matrix} \right\rangle (B + dA) \equiv \left\langle \begin{matrix} \end{matrix} \right\rangle (-A)^3$$

The state model and bracket polynomials introduced by L.H. Kauffman seem to be very special. They explore only the peculiar geometrical rules such as summation over the formal "ghost" degrees of freedom—all possible knot (link) splittings with simple defined weights. But one of the main advantages of the described construction is connected with the fact that Kauffman polynomials in A-value coincide with Jones knot invariants in t-value (where $t = A^{1/4}$).

Jones polynomial knot invariants were discovered first by V.F.R. Jones during his investigation of topological properties of braids (see Chapter 2 for details). Jones' proposition concerns the establishment of the deep connection between the braid group relations and the Yang-Baxter equations ensuring the necessary condition of transfer matrix commutativity [7]. The Yang-Baxted equations play an exceptionally important role in the statistical physics of integrable systems (such as ice, Potts, $O(n)$, 8-vertex, quantum Heisenberg models [24,30]).

The attempt to apply Kauffman invariants of regular isotopy for investigation of statistical properties of random walks with topological constraints in a thin slit has been made recently [31]. Below we extend the ideas of the work [31] considering the topological state of the knot as a special kind of *a quenched disorder.*

1.4. Lattice Knot Diagrams as Disordered Potts Model

Let us specify the model under consideration. Take a square lattice $\mathcal{M}$ turned to the angle $\pi/4$ with respect to the x-axis and project a knot embedded in $\mathbf{R}^3$ onto $\mathcal{M}$ supposing that each crossing point of the knot diagram coincides with one lattice vertex without fall (there are no empty lattice vertices)—see fig.1.6. Define the passages in all N vertices and choose such boundary conditions which ensure the lattice to form a single closed path; that is possible when $\sqrt{N}$ (i.e. N) is an odd number. The *frozen pattern* of all passages $\{b_i\}$ on the lattice together with the boundary conditions fully determine the topology of some 3D knot.

Of course, the model under consideration is rather rough because we neglect the "space" degrees of freedom due to trajectory fluctuations and keep the pure topological specificity of the system. Later on in Chapter 4 we discuss the applicability of such model for real physical systems and produce arguments in support of its validity.

The basic question of our interest is as follows: what is the probability $\mathcal{P}_N\{f[K]\}$ to find a knot diagram on our lattice $\mathcal{M}$ in a topological state characterized by some specific Kauffman invariant $f[K]$ among all 2^N microrealizations of the disorder $\{b_i\}$ in the lattice vertices. That probability distribution reads (compare to Eq.(1.1))

$$\mathcal{P}_N\{f[K]\} = \frac{1}{2^N} \sum_{\{b_i\}} \Delta \left[f[K\{b_1, b_2, \ldots, b_N\}] - f[K] \right] \tag{1.38}$$

where $f[K\{b_1, \ldots, b_N\}]$ is the representation of the Kauffman invariant as a function of all passages $\{b_i\}$ on the lattice $\mathcal{M}$. These passages can be regarded as a sort of quenched "external field" (see below).

Our main idea of dealing with Eq.(1.38) consists in two steps:

(a) At first we convert the Kauffman topological invariant into the known and well-investigated Potts spin system with the disorder in interaction constants;

(b) Then we apply the methods of the physics of disordered systems to the calculation of thermodynamic properties of the Potts model. It enables us to extract finally the estimation for the requested distribution function.

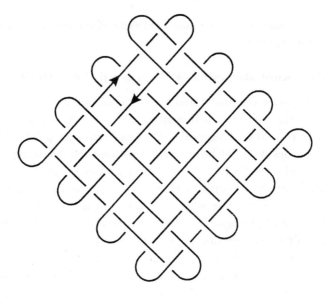

Fig. 1.6. Lattice knot with topological disorder realized in a quenched random pattern of passages.

Strictly speaking, we could have disregarded point (a), because it does not lead directly to the answer to our main problem. Nevertheless we follow the mentioned sequence of steps in pursuit of two goals:

– We would like to prove that the topologically-probabilistic problem can be solved within the framework of standard thermodynamic formalism;

– We would like to employ the knowledges accumulated already in physics of disordered Potts systems to avoid some unnecessary complications.

1.4.1. *Algebraic Invariants of Regular Isotopy*

To construct the Potts representation of the Kauffman polynomial invariant Eq.(1.29) of regular isotopy for some given pattern of "topological disorder", $\{b_i\}$, recall that any arbitrary set of simultaneous splittings in all lattice vertices represents the polygon decomposition of the lattice $\mathcal{M}$ which resembles a densely packed system of disjoint and non-selfintersecting circles. The collection of all polygons (circles) can be interpreted as a system of the so-called *Eulerian circuits* completely filling the square lattice.

Eulerian circuits are in one-to-one correspondence with the graph expansion of some disordered Potts system introduced in Section 1.3.1 (see details below and in [34]).

After these philosophical remark we are able to formulate the following theorem:

Theorem 3 (a) *Take N-vertex knot diagram on the lattice $\mathcal{M}$ with given boundary conditions and fixed set of passages $\{b_i\}$. (b) Take the dual lattice $\mathcal{L}$ in one-to-one corredpondence with $\mathcal{M}$ where one vertex of $\mathcal{M}$ belongs to one edge of $\mathcal{L}$.*

The Kauffman topological invariant $\langle K(A) \rangle$ of regular isotopy for knot diagrams on $\mathcal{M}$ admits representation in form of 2D Potts system on the dual lattice $\mathcal{L}$:

$$\langle K(A) \rangle = H\big(A, \{b_{kl}\}\big) \, Z_{potts}\big[q(A), \{J_{kl}(b_{kl}, A)\}\big] \qquad (1.39)$$

where:

$$H\big(A, \{b_{kl}\}\big) = \big(A^2 + A^{-2}\big)^{-(N+1)} \exp\left(\ln A \sum_{\{kl\}} b_{kl}\right) \qquad (1.40)$$

is the trivial multiplier (H does not depend on Potts spins);

$$Z_{potts}\big[q(A), \{J_{kl}(b_{kl}, A)\}\big] = \sum_{\{\sigma\}} \exp\left\{\sum_{\{kl\}} \frac{J_{kl}(b_{kl}, A)}{T} \delta(\sigma_k, \sigma_l)\right\} \qquad (1.41)$$

is the Potts partition function with interaction constants, J_{kl}, and number of spin states, q, defined as follows

$$\frac{J_{kl}}{T} = \ln[-A^{-4b_{kl}}]; \qquad q = \big(A^2 + A^{-2}\big)^2 \qquad (1.42)$$

and the variables b_{kl} play a role of disorder on edges of the lattice $\mathcal{L}$ dual to the lattice $\mathcal{M}$. The connection between b_{kl} and b_i is defined by convention

$$b_{kl} = \begin{cases} -b_i & \text{if } (kl)\text{-edge is vertical} \\ +b_i & \text{if } (kl)\text{-edge is horizontal} \end{cases} \qquad (1.43)$$

Proof. Rewrite the Kauffman invariant of regular isotopy, $\langle K \rangle$, in form of disordered Potts model defined in the previous section. Introduce the two-state "ghost" spin variables, $s_i = \pm 1$ in each lattice vertex independent on the crossing in the same vertex

$$\underset{\bigcap}{\bigcup} \quad s_i = +1 \qquad \text{and} \qquad)(\quad s_i = -1$$

Irrespective of the orientation of the knot diagram shown in fig.1.6 (i.e. restricting with the case of regular isotopic knots), we have

$$\langle K\{b_i\}\rangle = \sum_{\{S\}} \left(A^2 + A^{-2}\right)^{S-1} \exp\left(\ln A \sum_{i=1}^{N} b_i s_i\right) \qquad (1.44)$$

Written in such form the partition function $\langle K\{b_i\}\rangle$ represents the weihgted sum of all possible *Eulerian circuits* [†] on the lattice $\mathcal{M}$. Let us show explicitly that the microstates of the Kauffman system are in one-to-one correspondence with the microstates of some disordered Potts model on a lattice. Apparently for the first time the similar statement was expressed in the paper [8]. To be careful, we would like to use the following definitions:

(i) Let us introduce the lattice $\mathcal{L}$ dual to the lattice $\mathcal{M}$, or more precisely, one of two possible (odd and even) diagonal dual lattices, shown in fig.1.7. It can be easily noticed that the edges of the lattice $\mathcal{L}$ are in one-to-one correspondence with the vertices of the lattice $\mathcal{M}$. Thus, the disorder on the dual lattice $\mathcal{L}$ is determined on the *edges*. In turn, the edges of the lattice $\mathcal{L}$ can be divided into the subgroups of vertical and horisontal bonds. Each kl-bond of the lattice $\mathcal{L}$ carries the "disorder variable" b_{kl} being a function of the variable b_i located in the corresponding i-vertex of the lattice $\mathcal{M}$. The simplest and most sutable choice of the function $b_{kl}(b_i)$ is as in Eq.(1.43) (or vice versa for another choice of dual lattice); i is the vertex of the lattice $\mathcal{M}$ belonging to the kl-bond of the dual lattice $\mathcal{L}$.

(ii) For the given configuration of splittings on $\mathcal{M}$ and chosen dual lattice $\mathcal{L}$ let us accept the following convention: we mark the edge of the $\mathcal{L}$-lattice by the solid line if this edge is not intersected by some polygon on the $\mathcal{M}$-latice and we leave the corresponding edge unmarked if it is intersected by any polygon—as it is shown in the fig.1.7. Similarly, the sum $\sum s_i b_i$ in Eq.(1.44) can be rewritten in terms of marked and unmarked bonds on the

---

[†] Eulerian circuit is a trajectory on the graph which visits once and only once all graph edges.

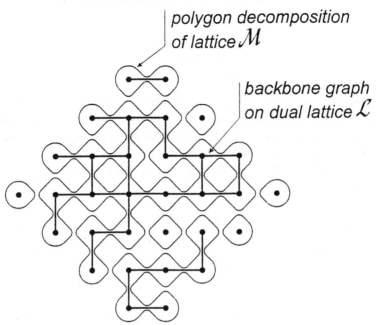

Fig. 1.7. Disintegration of the knot diagram on the $\mathcal{M}$-lattice into ensemble of nonselfintersecting loops (Eulerian circuits) and graph representation of the Potts model on the dual $\mathcal{L}$-lattice.

$\mathcal{L}$-lattice

$$
\begin{aligned}
\sum_i s_i b_i &= \underset{\text{mark}}{\sum} s_i b_i + \underset{\text{nonmark}}{\sum} s_i b_i \\[2mm]
&= \underset{\substack{\text{horiz}\\\text{mark}}}{\sum} s_i b_i + \underset{\substack{\text{vertic}\\\text{mark}}}{\sum} s_i b_i + \underset{\substack{\text{horiz}\\\text{nonmark}}}{\sum} s_i b_i + \underset{\substack{\text{vertic}\\\text{nonmark}}}{\sum} s_i b_i \\[2mm]
&= -\underset{\substack{\text{horiz}\\\text{mark}}}{\sum} b_{kl} - \underset{\substack{\text{vertic}\\\text{mark}}}{\sum} b_{kl} + \underset{\substack{\text{horiz}\\\text{nonmark}}}{\sum} b_{kl} + \underset{\substack{\text{vertic}\\\text{nonmark}}}{\sum} b_{kl} \\[2mm]
&= \underset{\text{nonmark}}{\sum} b_{kl} - \underset{\text{mark}}{\sum} b_{kl} = \underset{\text{all edges}}{\sum} b_{kl} - 2\underset{\text{mark}}{\sum} b_{kl}
\end{aligned}
$$

(1.45)

where we used the relation $\underset{\text{nonmark}}{\sum} b_{kl} + \underset{\text{mark}}{\sum} b_{kl} = \underset{\text{all edges}}{\sum} b_{kl}$.

(iii) Let m_s be the number of marked edges and C_s be the number of connected components of marked graph. Then the Euler relation reads:

$$S = 2C_s + m_s - N + \chi \tag{1.46}$$

The Eq.(1.46) can be proved directly. The χ-value depends on the genus of the surface, which can be covered by the given lattice, (i.e. χ depends on the boundary conditions). In the thermodynamic limit $N \gg 1$ the χ-dependence should disappear (at least for the flat surfaces), so the standard equality $S = 2C_s + m_s - N$ will be assumed below.

By means of definitions (i)-(iii), we can easily convert Eq.(1.44) into the form:

$$
\langle K\{b_{kl}\} \rangle = (A^2 + A^{-2})^{-(N+1)} \prod_{\substack{\text{all edges}}}^{N} [A^{b_{kl}}]
$$
$$
\times \sum_{\{G\}} (A^2 + A^{-2})^{2C_s} \prod_{\substack{\text{mark}}}^{m_s} [A^{-2b_{kl}}(-A^2 - A^{-2})] \tag{1.47}
$$

where we used Eq.(1.45) and the fact that $N + 1$ is even. Comparing Eq.(1.47) with Eq.(1.26) we immediately conclude that

$$
\sum_{\{G\}} (A^2 + A^{-2})^{2C_s} \prod_{\substack{\text{mark}}}^{m_s} [A^{-2b_{kl}}(-A^2 - A^{-2})] \equiv \sum_{\{\sigma\}} \prod_{\{kl\}} (1 + v_{kl}\delta(\sigma_k, \sigma_l)) \tag{1.48}
$$

what coincides with the partition function of the Potts system written in the form of dichromatic polynomial. Therefore, we have

$$
v_{kl} \overset{def}{=} A^{-2b_{kl}}(-A^2 - A^{-2}) = -1 - A^{-4b_{kl}}
$$
$$
q \overset{def}{=} (A^2 + A^{-2})^2 \tag{1.49}
$$

Since the "disorder" variables b_{kl} take the discrete values ± 1 only, we get the following expression for the interaction constant J_{kl} (see Eq.(1.27))

$$
\frac{J_{kl}}{T} = \ln\left[1 - (A^2 + A^{-2})A^{-2b_{kl}}\right] = \ln[-A^{-4b_{kl}}] \tag{1.50}
$$

Combining Eqs.(1.47)-(1.50) we obtain the statement of the Theorem 3 □.

Eq.(1.48) has the sense of partition function of the 2D disordered Potts system with the random nearest-neighbor interactions whose distribution remains arbitrary. The set of passages $\{b_{kl}\}$ uniquely determines the actual topological state of the woven carpet for the definite boundary conditions. Therefore the topological problem of the knot invariant determination is reduced to usual statistical problem of calculation of the partition function of the Potts model with the disorder in the interaction constants. Of course, this correspondence is still rather formal because the polynomial variable A is absolutely arbitrary and can take even complex values, but for some regions of A that thermodynamic analogy makes sense and could be useful as we shall see below.

The specific feature of the Potts partition function which gives the representation of the Kauffman algebraic invariant is connected with the existence of the relation between the temperature T and the number of spin states q (see Eq.(1.49)) according to which T and q cannot be considered as independent variables.

1.4.2. *Algebraic Invariants of Ambient Isotopy*

The representation of the algebraic topological invariant, $f[K]$, with respect to all Reidemeister moves (see Eq.(1.37)) for our system shown in the fig.1.6 is related to the oriented Eulerian circuits called *Hamiltonian walks* [†].

Let us suppose that the orientation of the knot diagram shown in fig.1.6 is choosen according to the natural orientation of the path representing a knot K in $\mathbf{R}^3$. For the defined boundary conditions we get the so-called *Manhattan lattice* consisting of woven threads with alternating directions.

It follows from the definition of twisting $Tw(K)$ (see the Section 1.3.2) that $Tw(K)$ changes the sign if the direction of one arrow in the vertex is changed to the inverse. Reversing the direction of any arrows in the given vertex even times we return the sign of twisting to the initial value.

We define groups of "even" and "odd" vertices on the lattice $\mathcal{M}$ as follows. The vertex i is called *even* (*odd*) if it belongs to the horizontal (vertical) bond (kl) of the dual lattice $\mathcal{L}$. Now it is easy to prove that the twisting of the knot on the Manhattan lattice $\mathcal{M}$ can be written in terms of above defined variables b_{kl}. Finally the expression for the algebraic

---

[†]A Hamiltonian walk is a closed path which visits once and only once all vertices of the given *oriented* graph.

invariant of ambient isotopy $f[K]$ on the lattice $\mathcal{L}$ reads

$$f[K] = \exp\left(3\ln[-A]\sum_{\{kl\}} b_{kl}\right)\langle K\left(\{b_{kl}\}, A\right)\rangle \qquad (1.51)$$

where $\langle K\left(\{b_{kl}\}, A\right)\rangle$ is defined by Eq.(1.39).

1.5. Annealed and Quenched Realizations of Topological Disorder

Fixed topological structure of a trajectory of given length fluctuating in space is a typical example of a quenched disorder. Actually, the knot structure is formed during the random closure of the path and cannot be changed without the path rupture. Because of the topological contsraints the entire phase space of ensemble of randomly generated closed loops is divided into the separated domains resembling the multi-valley structure of the spin glass phase space. Every domain corresponds to the sub-space of the path configurations with the fixed value of the topological invariant. The methods of theoretical description of the systems with quenched disorder in interaction constants are rather well developed, especially in regard to the investigation of *spin glass models* [25].

Central for these methods is the concept of *self-averaging* which can be explained as follows. Take some additive function F (the free energy, for instance) of some disordered spin system. The function F is the self-averaging quantity if the observed value, F_{obs}, of any macroscopic sample of the system coincides with the value F_{av} averaged over the ensemble of disorder realizations:

$$F_{obs} = \langle F\rangle_{av}$$

The phenomenon of self-averaging takes place in the systems with sufficiently weak long-range correlations: only in this case F can be considered as a sum of contributions from different volume domains, containing statistically independent realizations of disorder (for more details see [26]).

The central technical problem is in calculation of the free energy $F = -T\ln Z$ averaged over the randomly distributed quenched pattern in the interaction constants. In Section 1.4.2 we show that this famous thermodynamic problem of the spin glass physics is closely related to the knot entropy calculation.

Another problem arises when averaging the partition function Z (but not the free energy) over the disorder. Such problem is much simpler from computational point of view and corresponds to the case of *annealed* disorder. Usually the thermodynamic behavior of systems with annealed disorder is less rich and less interesting than that of systems with quenched disorder. However in our case it is necessary to consider the annealed random pattern $\{b_{kl}\}$, reflecting the situation when the topology of the closed loop can be changed. It means that the topological invariant, i.e. the Potts partition function, has to be averaged over all possible realizations of the pattern disorder. Such situation can be reached as follows: take ensemble of *open* paths in thermodynamic equilibrium and find the mean value of the topological invariants $\overline{\langle K(A) \rangle}$ or $\overline{f[K]}$ of a knot obtained by random path closure. As it is shown below, that calculation of the mean values of topological invariants allows to extract rather rough but nontrivial information about the knot statistics.

1.5.1. *Averaged Kauffman Invariants for Annealed Topological Disorder*

For annealed distribution the mean value of the topological invariant of regular isotopy, $\overline{\langle K(A) \rangle}$, reads

$$\overline{\langle K(A) \rangle} = \sum_{\{b_i\}} \langle K\{b_i, A\} \rangle \, \Theta\{b_i\} \tag{1.52}$$

where $\Theta\{b_i\}$ is the probability distribution of different realizations of the disorder pattern $\{b_i\}$ (i.e. of different microrealizations of the knot diagrams). The averaged invariant of ambient isotopic knots, $\overline{f[K(A)]}$ can be defined analogously:

$$\overline{f[K(A)]} = \sum_{\{b_i\}} f[K\{b_i, A\}] \, \Theta\{b_i\} \tag{1.52a}$$

In principle the distribution $\Theta\{b_i\}$ depends on statistics of the path in underlying 3D space and is determined physically by the process of the knot formation. Here we restrict ourselves to the following simplest suppositions:

(i) We regard crossings $\{b_i\}$ in different vertices of $\mathcal{M}$-lattice as completely uncorrelated variables (or, in other words, we assume that the variables $\{b_{kl}\}$ defined on the edges of the $\mathcal{L}$-lattice are statistically independent):

$$\Theta\{b_i\} = \prod_{i}^{N} P(b_i) \tag{1.53}$$

(ii) We suppose variable b_i (or b_{kl}) to take values ± 1 with equal probabilities, i.e.:

$$P(b_i) = \frac{1}{2}\,\delta(b_i - 1) + \frac{1}{2}\,\delta(b_i + 1) \qquad (1.54)$$

THE REGULAR ISOTOPY. Collect Eqs.(1.44) and (1.39) together with Eqs.(1.53)-(1.54) and substitute the result into Eq.(1.52). We get the following chain of relations

$$
\begin{aligned}
\overline{\langle K(A)\rangle} &= \sum_{\{S\}} \left(-A^2 - A^{-2}\right)^{S-1} \prod_{i}^{N} \left[A^{s_i b_i} \frac{1}{2}\Big(\delta(b_i - 1) + \delta(b_i + 1)\Big)\right] \\
&= \left(\frac{A + A^{-1}}{2}\right)^N \sum_{\{S\}} \left(-A^2 - A^{-2}\right)^{S-1} \\
&= \overline{H}_{reg}(A, N) \sum_{\{G\}} \left(-A^2 - A^{-2}\right)^{2C_s} \left(-A^2 - A^{-2}\right)^{m_s} \\
&= \overline{H}_{reg}(A, N) Z_{potts}^{crit}(A, N)
\end{aligned}
$$
$$(1.55)$$

where

$$\overline{H}_{reg}(A, N) = \left(A^2 + A^{-2}\right)^{-(N+1)} \cosh^N (\ln A) \qquad (1.56)$$

In Eq.(1.55) we took into account the possibility to change the orders in summations over splittings $\{S\}$ and disorder variables $\{b_i\}$. By $Z_{potts}^{crit}(A, N)$ we denote the partition function of the critical 2D-Potts model without any randomness and with the averaged interacton constant $\frac{J}{T}$ (compare to Eqs.(1.27)-(1.28)):

$$\frac{J}{T} = \ln\left(1 - A^2 - A^{-2}\right) \qquad (1.57)$$

Analysing Eqs.(1.55) and (1.57) we conclude that

$$\exp\left(\frac{J}{T}\right) - 1 = q^{1/2} \qquad (1.58)$$

This relation coincides with the *self-duality* condition of the 2D-Potts system [24] which ensures the model to be exactly solvable in the thermodynamic limit $N \to \infty$.

We have already mentioned that so far as the polynomial variable A takes completely arbitrary values, the partition function $Z_{potts}^{crit}(A, N)$

written in terms of the graph expansion in Eq.(1.55) is nothing else but an abbreviation. To attach physical sense to it, one should demand the parition function to converge. That is certainly fulfilled if

$$\left(A^2 + A^{-2}\right) = q \geq 0$$

$$\ln\left(1 - A^2 - A^{-2}\right) = \frac{J}{T} \geq 0 \tag{1.59}$$

These conditions are sufficient only and we search their solution in the form $A = e^u e^{i\left(\alpha + \frac{\pi}{2}\right)}$ (u is positive), covering two very important cases when the considered thermodynamic language is applicable:

(a) $u = 0$ (i.e. the polynomial variable A is the root of unity); the allowed values of α are

$$-\frac{\pi}{4} < \alpha \leq \frac{\pi}{4} \quad \rightarrow \quad \begin{cases} 0 \leq q = 4\cos^2 2\alpha \leq 4 \\[2mm] \dfrac{J}{T} = \ln(1 + 2\cos 2\alpha) \end{cases} \tag{1.60}$$

(b) $\alpha = 0$; the allowed values of u are

$$0 < u < \infty \quad \rightarrow \quad \begin{cases} q = 4\cosh^2 2u > 4 \\[2mm] \dfrac{J}{T} = \ln(1 + 2\cosh 2u) \end{cases} \tag{1.61}$$

The free energy of the Potts model on the square lattice at the critical line (1.58) has been calculated in the limit $N \to \infty$ by means of Bethe-ansatz [24]. We reproduce below the final results of these calculations for two cases defined in Eqs.(1.60) and (1.61):

$$\lim_{N \to \infty} \frac{1}{N} \ln Z_{potts}^{crit}(A, N) = \frac{1}{2}\ln q + I(q) \tag{1.62}$$

where

(i) $0 \leq q \leq 4$; $2\cos 2\alpha = q^{1/2}$ (see case (a))

$$I(q) = \int_{-\infty}^{\infty} \frac{dx}{x} \tanh 2\alpha x \frac{\sinh(\pi - 2\alpha)x}{\sinh \pi x}; \tag{1.63}$$

(ii) $q > 4$; $2\cosh 2u = q^{1/2}$ (see case (b))

$$I(q) = 2u + 2\sum_{n=1}^{\infty} n^{-1} e^{-2nu} \tanh 2nu \tag{1.64}$$

Thus, the ensemble of knots uniformly distributed on the lattice $\mathcal{M}$ is characterized by the following mean values of the topological invariant in the thermodynamic limit

$$
\lim_{N \to \infty} \frac{1}{N} \ln \overline{\langle K(A) \rangle} = \lim_{N \to \infty} \frac{1}{N} \ln \overline{H}_{reg}(A, N) + \frac{1}{2} \ln q + I(q)
$$

$$
= \begin{cases} i\pi + \ln \cos \alpha + I(q) & \text{in case (a)} \\ i\frac{\pi}{2} + \ln \sinh u + I(q) & \text{in case (b)} \end{cases} \qquad (1.65)
$$

In particular we have for $u = 0$ $\alpha = \frac{\pi}{4}$ (i.e. q=0)

$$
\lim_{N \to \infty} \frac{1}{N} \ln \overline{\langle K \left(A = ie^{i\frac{\pi}{4}} \right) \rangle} = i\pi - \frac{1}{2} \ln 2 + \frac{4}{\pi} \mathbf{G} \qquad (1.66)
$$

where $\mathbf{G} = 1 - 3^{-2} + 5^{-2} - \ldots$ is the Catalan's constant.

THE AMBIENT ISOTOPY. The average value of the invariant $f[K(A)]$ of ambient isotopic knots can be calculated using the Potts representation derived in Section 1.4.2. We have from Eqs.(1.51) and (1.52a)

$$
\overline{f[K(A)]} = \left(A^2 + A^{-2} \right)^{-(N+1)} \sum_{\{b_{kl}\}} \Theta\{b_{kl}\} \times
$$

$$
\sum_{\{\sigma\}} \prod_{\{kl\}} \exp \left\{ \left(3 \ln[-A] + \ln A \right) b_{kl} + \frac{J_{kl}}{T} \delta(\sigma_k, \sigma_l) \right\}
$$

$$
(1.67)
$$

After some simple algebra we get

$$
\overline{f[K(A)]} = - \left(\frac{A^4 + A^{-4}}{2} \right)^N \left(A^2 + A^{-2} \right)^{-(N+1)} \sum_{\{\sigma\}} \prod_{\{kl\}} \exp \left\{ \frac{\overline{J}}{T} \delta(\sigma_k, \sigma_l) \right\}
$$

$$
(1.68)
$$

where the average interaction constant $\frac{\overline{J}}{T}$ is defined as follows

$$
\frac{\overline{J}}{T} = \ln \left(\frac{A^4 + A^{-4}}{2} \right) = -\ln \left(1 - \frac{q}{2} \right) \qquad (1.69)
$$

The Potts system defined by Eqs.(1.68)-(1.69) is off-critical and in general its free energy cannot be computed exactly even in the thermodynamic limit. However it is practicable when Eq.(1.69) is consistent with the self-duality condition (1.58) for the Potts system. Thus, the requested A-values are the roots of the system

$$\begin{cases} \dfrac{\overline{J}}{T} = -\ln\left(\dfrac{A^4 + A^{-4}}{2}\right) \\[4mm] \dfrac{\overline{J}}{T} = \ln\left(1 - A^2 - A^{-2}\right) \end{cases} \tag{1.70}$$

The solution of (1.70) gives only one critical value $A = e^{i\frac{\pi}{2}}$, i.e. $q = 4$. At this point the averaged value of the ambient isotopic Kauffman invariant in the thermodynamic limit is

$$\lim_{N\to\infty} \frac{1}{N} \ln \overline{f\left[K\left(A = e^{i\frac{\pi}{2}}\right)\right]} = \lim_{q\to 4} I(q) = -2 \tag{1.71}$$

The reasonable estimation for averaged topological invariant $\overline{f[K(A)]}$ (Eq.(1.68)) can be obtained within the framework of the mean-field approximation. We leave these computations for exercise. The examples of mean-field approach in context of topological problems see in the next section where we calculate the knot entropy for our specific lattice model.

1.5.2. *Entropy of Knots. Replica Methods*

We would like to proceed with the calculation of the probability distribution $\mathcal{P}_N\{f[K]\}$ (see Eq.(1.38)). Although we are unable to evaluate this function exactly, the representation of $\mathcal{P}_N\{f[K]\}$ in terms of disordered Potts system enable us to give an upper estimation for the fraction of randomly generated paths belonging to some definite topological class (in particular, to the trivial one). We use the following chain of inequalities restricting ourselves with the case of regular isotopic knots for simplicity ([31]):

Probability $\mathcal{P}_N$ of knot formation in a given topological state	$\leq$	Probability $\mathcal{P}_N\{K(A)\}$ of knot formation with specific topological invariant $\langle K(A)\rangle$ for *all* A

	$\leq$	Probability $\mathcal{P}_N\{K(A^*)\}$ of knot formation for *specific* value of A^* minimizing the free energy of associated Potts system

$$\tag{1.72}$$

The first inequality is due to the fact that Kauffman invariant of regular isotopic knots is not a complete topological invariant, whereas the last probability in the chain can be written as follows

$$P_N\{K(A^*)\} = \sum_{\{b_{kl}\}} \Theta\{b_{kl}\}\Delta\Big[\langle K\{b_{kl}, A^*\}\rangle - \langle K(A^*)\rangle\Big] \qquad (1.73)$$

where $\sum$ means summation over all possible configurations of the "crossing field" $\{b_{kl}\}$, Δ-function cuts out all states of the field $\{b_{kl}\}$ with specific value of Kauffman invariant $\langle K\{b_{kl}, A^*\}\rangle \equiv \langle K(A^*)\rangle$ and $\Theta\{b_{kl}\}$ is the probability of realization of given crossings configuration. The probability of trivial knot formation can be estimated in the similar way

$$\begin{aligned} P_N^{(0)}(A^*) &\leq \sum_{\{b_{kl}\}} \Theta\{b_{kl}\}\Delta\Big[\ln\langle K\{b_{kl}, A^*\}\rangle\Big] \\ &\simeq \frac{1}{2\pi}\int_{-\infty}^{\infty} dy \int \cdots \int \prod_{kl} P(b_{kl}) db_{kl} \langle K^{iy}\{b_{kl}, A^*\}\rangle \end{aligned} \qquad (1.74)$$

where $\langle K(A^*)\rangle \equiv 1$ for trivial knots.

Thus our problem is reduced to the calculation of non-integer complex moments of the partition function, i.e., values of the type $\overline{\langle K^{iy}\{b_{kl}, A^*\}\rangle}$. An analogous problem of evaluation of non-integer moments is well known in the spin-glass theory. Indeed, the averaging of the free energy of the system, $\overline{F}$, over quenched random field is widely performed via so-called *replica-trick* [36]. The idea of the method is as follows. Consider the identity $Z^n \equiv e^{n \ln Z}$ and expand the right-hand side up to the first order in n. We get $Z^n = 1 + n\ln Z + O(n^2)$. Now we can write

$$F = -\ln Z = -\lim_{n\to 0} \frac{Z^n - 1}{n}$$

Thus instead of dealing with the logarithm one have to average the powers ("moments") of the partition function. Suppose now that it is possible to perform averaging $\overline{Z^n}$ for arbitrary *positive integer* value of n and take the limit $n \to 0$ at the end of all calculations. Why such method should work properly? There are rather obscure speculations as well as some verifications of the replica approach by the supersymmetry methods, but apparently the strongest argument in support of the validity of the replica approach is the fact that in most cases it works well for various systems.

We proceed with the calculation of the complex moments of the partition function $\langle K\{b_{kl}\}\rangle$. In other words we would like to find the averaged value $\overline{\langle K^n \rangle}$ for integer values of n. Then we put $n = iy$ and compute the remaining integral in Eq.(1.74) over y-value. Of course, this procedure needs to be verified. Although our approach is no more curious than replica one, it would be extremely desirable to test the results obtained by means of computer simulations.

The outline of our calculations is as follows. We begin by rewriting the averaged Kauffman invariant using the standard representation of the replicated Potts partition function and extract the corresponding mean-field free energy $\overline{F(A)}$. Minimizing $\overline{F(A)}$ with respect to A we find the equilibrium value A^*. Then we compute the desired probability of trivial knot formation $\mathcal{P}_N^{(0)}(A^*)$ evaluating the remaining Gaussian integrals.

Averaging the nth power of Kauffman invariant over independent values of the "crossing field" $b_{kl} = \pm 1$ we get

$$
\overline{\langle K^n(A) \rangle} = \int \cdots \int \prod_{kl} P(b_{kl}) db_{kl} K^{2n}\{b_{kl}\}
$$

$$
= [2\cosh(2\beta)]^{-2n(N+1)}
$$

$$
\times \sum_{\{\sigma\}} \prod_{kl} \exp\left\{ i\pi \sum_{kl} \delta(\sigma_k^\alpha, \sigma_l^\alpha) + \ln\cosh\left[\beta \sum_{\alpha=1}^{n} \left(4\delta(\sigma_k^\alpha, \sigma_l^\alpha) - 1 \right) \right] \right\}
$$

$$
\tag{1.75}
$$

where $\beta = \ln A$. Let us break for a moment the connection between the number of spin states, q, and interaction constant and suppose $|\beta| \ll 1$. Later on we shall verify the selfconsistency of this approximation. Now the exponent in the last expression can be expanded as a power series in β. Keeping the terms of order β^2 only, we rewrite Eq.(1.75) in the standard form of n-replica Potts partition function

$$
\overline{\langle K^n(A) \rangle} = [2\cosh(2\beta)]^{-2n(N+1)} \exp\left[N\left(\frac{1}{2}\beta^2 n^2 \right) \right]
$$

$$
\times \sum_{\{\sigma_1 \ldots \sigma_n\}} \exp\left\{ \frac{J^2}{2} \sum_{kl}^{N} \sum_{\alpha \neq \beta}^{n} \sigma_{ka}^\alpha \sigma_{kb}^\beta \sigma_{la}^\alpha \sigma_{lb}^\beta \right.
$$

$$
\left. + \left(\frac{J^2}{2}(q-2) + \bar{J}_0 \right) \sum_{kl}^{N} \sum_{\alpha=1}^{n} \sigma_{ka}^\alpha \sigma_{lb}^\beta \right\}
$$

$$
\tag{1.76}
$$

where spin indexes a, b change in the interval $[0, q-1]$, $\beta^2 \ll 1$ and

$$
\begin{aligned}
J^2 &= 16\beta^2 \\
\bar{J}_0 &= i\pi - 4\beta^2 n \\
q &= 4 + 16\beta^2 > 4
\end{aligned}
\tag{1.77}
$$

According to the results of Cwilich and Kirkpatrick [37] and later works (see, for instance, [39]), the spin-glass ordering takes place and the usual ferromagnetic phase makes no essential contribution to the free energy under the condition

$$
\frac{\bar{J}_0}{J} < \frac{q-4}{2}
\tag{1.78}
$$

Substituting Eq.(1.77) into Eq.(1.78) it can be seen that $\Re(\text{l.h.s.}) < \Re(\text{r.h.s})$ in Eq.(1.78) for all β. Thus, we expect that the spin-glass ordering (in the infinite-range model) corresponds to the solutions

$$
\begin{aligned}
m_a^\alpha &= \Big\langle q\delta(\sigma_k^\alpha, a) - 1 \Big\rangle = 0 \\
Q_{ab}^{\alpha\beta} &= \Big\langle q\delta(\sigma_k^\alpha, a) - 1 \Big\rangle \Big\langle q\delta(\sigma_k^\beta, b) - 1 \Big\rangle \neq 0
\end{aligned}
$$

where m_a^α and $Q_{ab}^{\alpha\beta}$ are the ferromagnetic and spin-glass order parameters respectively. If it is so, we can keep the term in the exponent (Eq.(1.76)) corresponding to interreplica interactions only.

We follow now the standard scheme of analysis of Potts spin glasses partition function exhaustively described in [37,38,39]; main steps of this analysis are shortly represented below. Performing the Hubbard-Stratonovich transformation to the scalar fields $Q_{iab}^{\alpha\beta}$ and implying the homogeneous isotropic solution of the form $Q_{iab}^{\alpha\beta} = Q_i^{\alpha\beta}\delta_{ab}$, we can write down the value $\overline{\langle K^n \rangle}$ (Eq.(1.76)) as follows ([37]):

$$
\overline{\langle K^n \rangle} = \exp\left\{ N\left[\ln \frac{\pi}{J^2} n(n-1)(q-1)^2 - \ln\left(2\cosh\frac{J}{2}\right) + \frac{J^2 n^2}{32} \right] \right\}
$$

$$
\times \sum_{\{\sigma\}} \int \prod_i dQ_i^{\alpha\beta} \exp\left\{ -\int H\{Q_i^{\alpha\beta}\} d^2x \right\}
\tag{1.79}
$$

where

$$H\{Q^{\alpha\beta}\} = (q-1)\left[\frac{1}{4}\left(\frac{2}{J^2}-1\right)\sum_{\alpha\neq\beta}(Q^{\alpha\beta})^2 - \frac{1}{6}\sum_{\alpha\neq\beta\neq\gamma}Q^{\alpha\beta}Q^{\beta\gamma}Q^{\gamma\alpha}\right.$$

$$-\frac{q-2}{12}\sum_{\alpha\neq\beta}(Q^{\alpha\beta})^3 - \frac{q-2}{4}\sum_{\alpha\neq\beta\neq\gamma}(Q^{\alpha\beta})^2 Q^{\beta\gamma}Q^{\gamma\alpha}$$

$$\left.-\frac{1}{8}\sum_{\alpha\neq\beta\neq\gamma\neq\delta}Q^{\alpha\beta}Q^{\beta\gamma}Q^{\gamma\delta}Q^{\delta\alpha} - \frac{q^2-6q+6}{48}\sum_{\alpha\beta}(Q^{\alpha\beta})^4\right]$$

$$(1.80)$$

In [38,37] it was shown, that the mean-field replica symmetric solution of the mean-field Potts spin glass is unstable for $q \geq 2$ and the right ansatz of Eqs.(1.79)-(1.80) corresponds to the first level of Parisi replica breaking scheme for spin glasses. Hence, we have

$$Q^{\alpha\beta} = \begin{cases} Q & \text{if } \alpha \text{ and } \beta \text{ belong to the same group of } m \text{ replicas} \\ 0 & \text{otherwise} \end{cases} \qquad (1.81)$$

Analysis shows that for $q > 4$ (our case) the transition to the glassy state corresponds to $m = 1$ which implies the accessory condition $F_{pm} = F_{sg}$, where F_{pm} and F_{sg} are the free energies of paramagnetic and spin-glass phases respectively. The transition occurs at the point

$$1 - \frac{2}{J^2} = \frac{(q-4)^2}{3(q^2 - 18q + 42)} \qquad (1.82)$$

Substituting Eq.(1.77) into Eq.(1.82) we find the self-consistent value of reverse temperature of a spin-glass transition, β_{tr}:

$$\beta_{tr} = 0.35 \qquad (1.83)$$

This numerical value is consistent with the condition $\beta_{tr}^2 \ll 1$ implied above in the course of expansion of Eq.(1.76).

According to the results of the work [37] the n-replica free energy near the transition point has the following form

$$F \simeq \frac{1}{64}Nn(q-1)^2 Q_{tr}\left(\frac{1}{\beta^2} - \frac{1}{\beta_{tr}^2}\right)^2 \qquad (1.84)$$

with the following expression of the spin-glass order parameter

$$Q_{tr} = \frac{2(4-q)}{q^2 - 18q + 42} > 0 \qquad (1.85)$$

From Eq.(1.84) we conclude that the free energy $\overline{F}$ reaches its minimum as a function of $A = \exp(\beta)$ just at the point $A^* = \exp(\beta_{tr})$. Using Eqs.(1.84) and (1.85) we rewrite the expression for the averaged n-replica Kauffman invariant $\overline{\langle K^n \rangle}$ in the vicinity of β_{tr} as follows (compare to [37]):

$$\overline{\langle K^{2n} \rangle} \simeq \exp\left\{ Nn^2\left[\left(3 + 16\beta^2\right)^2 \ln\frac{\pi}{16\beta^2} + \frac{\beta^2}{2} \right] \right.$$
$$- Nn\left[\left(3 + 16\beta^2\right)^2 \ln\frac{\pi}{16\beta^2} + \ln 2 + \frac{\beta^2}{2} \right. \qquad (1.86)$$
$$\left. \left. - \frac{\left(3 + 16\beta^2\right)^2 \left(\beta^{-2} - \beta_{tr}^{-2}\right)^2 \beta_{tr}^2}{\left(4 + 16\beta_{tr}^2\right)^2 - 18\left(4 + 16\beta_{tr}^2\right) + 42} \right] \right\}$$

Substituting Eq.(1.86) into Eq.(1.74) and bearing in mind, that $n = iy$, we can easily evaluate the remaining Gaussian integral over y-value and obtain the result for $\mathcal{P}_N^{(0)}(A)$. As it has been mentioned above, to get the simplest estimation for probability of trivial knot formation, we use the last inequality in the chain of equations (1.72) corresponding to the choice $A = A^* \equiv \exp(\beta_{tr})$:

$$\mathcal{P}_N^{(0)}(A^*) \simeq \exp(-1.04N) \qquad (1.87)$$

1.6. Remarks and Conclusions

Let us summarize briefly the advantages, disadvantages and perspectives of results obtained in Chapter 1.

1. Calculations of the averaged value of algebraic knot invariants over the random uncorrelated distribution of crossings enable us to extract some rough notion concerning the topological state of linear (open) chain in thermodynamic equilibrium. Let us define the power η_{av} of the averaged Kauffman invariant $\overline{\langle K \rangle}$ as follows:

$$\eta_{av} = \lim_{|A| \to \infty} \frac{\ln \overline{\langle K \rangle}}{\ln A} \qquad (1.88)$$

We will call η_{av} "complexity of average knot" (see Chapter 2 for more details). The results of Section 1.5 enable to conclude that $\eta_{av} \sim N$, which it is not surprising from the point of view of statistical mechanics because η_{av} is proportional to the free energy of the Potts system. But from the other hand we find remarkable the fact that the power of the average algebraic polynomial invariant η_{av} grows linearly with N. It means that the maximum of the distribution function $P(\eta, N)$ is in the region of very "complex" knots, i.e., knots far from trivial.

2. We show that estimation of the fraction of trivial knots can be considered within the framework of standard mean-field spin-glass theory. Our computations are oversimplified in many points discussed along the way (see Section 1.5.2) what makes the numerical value of the constant in Eq.(1.87) rather doubtfull. However we believe in selfconsistent calculations of the parameter A^* (see Eq.(1.83) and think that it gives the right value which minimizes the free energy of corresponding Potts system.

The results presented above could be generalized and improved in the following ways:

- Similar computations should be performed for estimation of fraction of trivial knots using the Kauffman invariant of ambient isotopy and the "complexity" of the quenched ensemble of randomly generated knots, η_q, needs to be calculated.

- An attempt should be made to receive an exact solution for knot entropy problem using the conformal methods applied for 2D random bond Potts systems [40].

- One have to reconsider the problem of knot entropy estimation using the Alexander polynomials as topological invariants. Here the path integral representation of Alexander invariants in terms of fermionic fields could be extremely useful.

- It is desirable to generalize our approach to the knot entropy calculation in the case of knot diagrams represented by random (but not regular) lattices.

References

1. *Fractional Statistics and Anyon Superconductivity* (WSPC: Singapore, 1990)
2. S. Majid, Int. J. Mod. Phys. (A) 8 (1993), 4521
3. C. Gómez, G. Sierra, J. Math. Phys., 34 (1993), 2119
4. V.F.R. Jones, Pacific J. Math., 137 (1989), 311
5. W.B.R. Likorish, Bull. London Math. Soc., 20 (1988), 558
6. M. Wadati, T.K. Deguchi, Y. Akutso, Phys. Rep., 180 (1989), 247
7. *Integrable Models and Strings*, Lect. Not. Phys., 436, (Springer: Heidelberg, 1994)
8. L.H. Kauffman, Topology, 26 (1987), 395.
9. L.H. Kauffman, AMS Contemp. Math. Series, 78 (1989), 263
10. A.V. Vologodskii, A.V. Lukashin, M.D. Frank-Kamenetskii, V.V.Anshelevich, Zh. Exp. Teor. Fiz., 66 (1974), 2153; Zh. Exp. Teor. Fiz., 67 (1974), 1875 (in Russian); M.D.Frank-Kamenetskii, A.V.Lukashin, A.V.Vologodskii, Nature, 258 (1975), 398
11. V.A. Vassiliev, *Complements of Discriminants of Smooth Maps: Topology and Applications*, Math. Monographs, 98 (Trans. AMS, 1992); D. Bar-Natan, preprint: Harvard, 1992
12. S.F. Edwards, Proc. Roy. Soc., 91 (1967), 513
13. M. Yor, J. Appl. Prob., 29 (1992), 202; Math. Finance, 3 (1993), 231
14. A. Comtet, J. Desbois, C. Monthus, J. Stat. Phys., 73 (1993), 433
15. F. Wiegel *Introduction to Path-Integrals Methods in Physics and Polymer Science* (WSPC: Singapore, 1986)
16. D.S. Khandekar, F. Wiegel, J. Phys. (A), 21 (1988), L-563; J. Phys. (Paris), 50 (1989), 2205
17. D. Hofstadter, Phys. Rev. (B), 14 (1976), 2239; J. Guillement, B. Helffer, P. Treton, J. Phys. (Paris), 50 (1989), 2019
18. J. Desbois, J.Phys. (A): Math. Gen., 25 (1992), L-195, L-755; J. Desbois, A. Comtet, J.Phys. (A): Math. Gen., 25 (1992), 3097
19. P.B. Wiegmann, A.V. Zabrodin, Phys. Rev. Lett., 72 (1994), 1890; Nucl. Phys. (B), 422 [FS] (1994), 495
20. J. Bellissard, Lect. Notes Phys., 257 (1986), 99
21. B.A. Dubrovin, S.P. Novikov, A.T. Fomenko, *Modern Geometry*, (Nauka: Moscow, 1979)
22. K. Iwata, S.F. Edwards, J. Chem. Phys., 90 (1989), 4567
23. F.Y. Wu, Rev. Mod. Phys., 54 (1982), 235
24. R.J. Baxter, *Exactly Solvable Models in Statistical Mechanics* (Academic Press: London, 1982)
25. Mezard, Parisi, Virasoro, *Spin Glass Theory and Beyond*, (World Scientific: Singapore, 1987)
26. I.M. Lifshits, S.A. Gredeskul, L.A. Pastur, *Introduction to Theory of Disordered Systems*, (Nauka: Moscow, 1982)

27. W.T. Tuttle, J. Combinatorial Theory, 2 (1967), 301
28. K. Reidemeister, *Knotentheorie* (Springer: Berlin, 1932)
29. L.H. Kauffman, *Knots and Physics* (WSPC: Singapore, 1991)
30. Lect. Notes Phys., 436, *Integrable Models and Strings* (Springer: Heidelberg, 1994)
31. A.Yu. Grosberg, S. Nechaev, J.Phys. (A): Math. Gen., 25 (1992), 4659
32. F.Y. Wu, J. Knot Theory Ramific., 1 (1992),
33. A.Yu. Grosberg, S. Nechaev, Europhys. Lett., 20 (1992), 613
34. B. Duplantier, F. David, J. Stat. Phys., 51 (1988), 327
35. F. Wu, Rev. Mod. Phys.,
36. S.F. Edwards, P.W. Anderson, J. Phys.(F), 5 (1975), 965
37. G. Cwilich, T.R. Kirkpatrick, J.Phys.A, 22 (1989), 4971
38. D.J. Gross, I. Kanter, H. Sompolinsky, Phys. Rev. Lett., 55 (1985), 304
39. E. De Santis, G. Parisi, F. Ritort, preprint: cond-mat/9410093
40. Vl. Dotsenko, M. Picco, P. Pujol, preprint: HEP-TH/9405003; preprint: HEP-TH/9501017; Vik.S. Dotsenko, Vl.S. Dotsenko, M. Picco, P. Pujol, preprint: HEP-TH/9502131

CHAPTER 2

RANDOM WALKS ON LOCAL NONCOMMUTATIVE GROUPS

2.1. Introduction

In the present Chapter we continue analysing the statistical problems in knot theory, but our attention is paid to some more delicate matters related to investigation of correlations in knotted random paths caused by the topological constraints. The methods elaborated in Chapter 1 allow us to discuss these questions but we find it more reasonable to take a look at the problems of knot entropy estimation in terms of conventional random matrix theory. Our consideration is based mainly on recent joint work with A.M. Vershik and A.Yu. Grosberg [1].

Our main aim here is as follows: we show that many non-trivial properties of the limit behavior of knot statistics can be explained in context of the limit behavior of random walks over the elements of some nonabelian (hyperbolic) group related to the so-called *braid* representation of knots (see Section 2.2 for details).

Another reason for us to consider the limit distributions (and conditional limit distributions) of Markov chains on *locally noncommutative discrete groups* [§] is due to the fact that this class of problems could be regarded as the first step in a consistent harmonic analysis on the multi-connected manifolds (like Teichmüller space). The simplest examples of that connection are discussed in Chapter 3.

[§]Notation introduced by A.M.Vershik.

2.2. Brownian Bridges on Simplest Noncommutative Groups and Knot Statistics

Investigation of the limit distributions of random walks on noncommutative groups is not a new subject in the probability theory. Namely, a set of rigorous results concerning the limit behavior of Markov chains on the free group and on the Riemann surface of the constant negative curvature has been received in papers [2,3,4]; the problem of construction of probability measure for the random walks on the modular group has been studied in the work [5]. To this theme we could also attribute a number of spectral problems considered in the theory of dynamic systems on the hyperbolic manifolds [6,7] as well as the subject of the random matrix theory [8].

However in the context of "topologically-probabilistic" consideration the problems dealing with distributions of noncommutative random walks are practically out of discussion, except for very few special cases [9,10,11,14,15]. Particularly, in these works it has been shown that statistics of random walks with the fixed topological state with respect to the regular array of obstacles on the plane can be obtained from the limit distribution of the so-called "brownian bridges" (see the definition below) on the universal covering—the graph with the topology of Cayley tree. The analytic construction of nonabelian topological invariant for the trajectories on the double punctured plane and statistics of simplest nontrivial random braid B_3 was shortly discussed in [12] (see Chapter 3 for details).

Below we calculate the conditional limit distributions of the brownian bridges on the braid group B_3 and derive the limit distribution of powers of Alexander polynomial of knots generated by random B_3-braids. We also discuss the limit distribution of random walks on locally free groups and express some conjectures about statistics of random walks on the group B_n. More extended discussion of the results concerning the statistics of Markov chains on the braid and locally free groups one can find in [13].

2.2.1. *Basic Definitions and Statistical Model*

Recall some necessary information concerning the definition of braid groups and construction of the algebraic knot invariants from the braid group representation.

BRAIDS. The braid group B_n of n strings has $n-1$ generators $\{\sigma_1, \sigma_2, \ldots,$

$\sigma_{n-1}\}$ with the following relations:

$$\sigma_i\sigma_{i+1}\sigma_i = \sigma_{i+1}\sigma_i\sigma_{i+1} \qquad (1 \le i < n-1)$$
$$\sigma_i\sigma_j = \sigma_j\sigma_i \qquad (|i-j| \ge 2) \qquad (2.1)$$
$$\sigma_i\sigma_i^{-1} = \sigma_i^{-1}\sigma_i = e$$

Let us mention that:

- Any arbitrary word written in terms of "letters"—generators from the set $\{\sigma_1,\ldots,\sigma_{n-1},\sigma_1^{-1},\ldots,\sigma_{n-1}^{-1}\}$—gives a particular *braid*. The geometrical interpretation of braid generators is shown in fig.2.1.

- The *length* of the braid is the total number of the used letters, while the *minimal irreducible length* hereafter referred to as the "primitive word" is the shortest noncontractible length of a particular braid which remains after applying all possible group relations Eq.(2.1). Diagramatically the braid can be represented as a set of crossed strings going from the top to the bottom (see fig.2.2) appeared after subsequent gluing the braid generators (fig.2.1).

- The closed braid appears after gluing the "upper" and the "lower" free ends of the braid on the cylinder.

- Any braid corresponds to some knot or link. So, it is feasible principal possibility to use the braid group representation for the construction of topological invariants of knots and links. However the correspondence between braids and knots is not mutually single valued and each knot or link can be represented by infinite series of different braids. This fact should be taken into account in course of knot invariant construction.

ALGEBRAIC INVARIANTS OF KNOTS. Take a knot diagram K in general position on the plane. Let $f[K]$ be the topological invariant of the knot K. One of the ways to construct the knot invariant using the braid group representation is as follows.

1. Represent the knot by some braid $b \in B_n$. Take the function f

$$f: \ B_n \to \mathbf{C}$$

Demand f to take the same value for all braids b representing the given knot K. That condition is established in the well-known theorem (see, for instance, [16]):

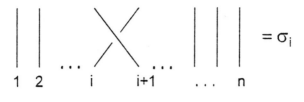

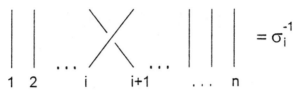

Fig. 2.1. Graphic representation of generators σ_i ("positive") and σ_i^{-1} ("negative") in the group B_n.

Theorem 4 (Markov-Birman) *The function $f_K\{b\}$ defined on the braid $b \in B_n$ is the topological invariant of a knot or link if and only if it satisfies the following "Markov condition":*

$$f_K\{b'\,b''\} = f_K\{b''\,b'\}$$
$$f_K\{b'\,\sigma_n\} = f_K\{\sigma_n\,b'\} = f_K\{b'\} \qquad b', b'' \in B_n \tag{2.2}$$

where b' and b'' are two subsequent subwords in the braid — see fig.2.3.

2. Now the invariant $f_K\{b\}$ can be constructed using the linear functional $\varphi\{b\}$ defined on the braid group and called *Markov trace*. It has the following properties

$$\varphi\{b'\,b''\} = \varphi\{b''\,b'\}$$
$$\varphi\{b'\,\sigma_n\} = \tau\varphi\{b'\} \tag{2.3}$$
$$\varphi\{b'\,\sigma_n^{-1}\} = \bar{\tau}\varphi\{b'\}$$

where

$$\tau = \varphi\{\sigma_i\}, \quad \bar{\tau} = \varphi\{\sigma_i^{-1}\}; \qquad i \in [1, n-1] \tag{2.4}$$

The invariant $f_K\{b\}$ of the knot K is connected with the linear functional $\varphi\{b\}$ defined on the braid b as follows

$$f_K\{b\} = (\tau\bar{\tau})^{-(n-1)/2} \left(\frac{\bar{\tau}}{\tau}\right)^{\frac{1}{2}\left(\#(+)-\#(-)\right)} \varphi\{b\} \tag{2.5}$$

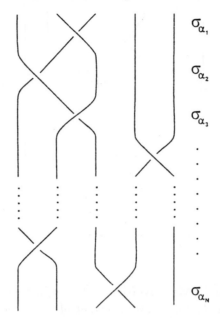

Fig. 2.2. Schematic representation of particular braid of N generators.

where $\#(+)$ and $\#(-)$ are the numbers of "positive" and "negative" cross-ings in the given braid correspondingly (see the fig.2.1).

The Alexander algebraic polynomials are the first well-known invari-ants of such type. In the beginning of 1980s Jones discovered the new knot invariants. He used the braid representation "passed through" the Hecke algebra relations, where the Hecke algebra, $H_n(t)$, for B_n satisfies both braid group relations Eq.(2.1) and an additional "reduction" relation (see the works [16,17])

$$\sigma_i^2 = (1-t)\sigma_i + t \tag{2.6}$$

Now the trace $\varphi\{b\} = \varphi(t)\{b\}$ can be said to take the value in the ring of polynomials of one complex variable t. Consider the functional $\varphi(t)$ over the braid $\{b'\,\sigma_i\,b''\}$. Eq.(2.6) allows us to get the recursion (skein) relations for $\varphi(t)$ and for the invariant $f_K(t)$ (see for details [20]):

$$\varphi(t)\{b'\sigma_i b''\} = (1-t)\varphi(t)\{b'b''\} + t\varphi(t)\{b'\sigma_i^{-1}b''\} \tag{2.7}$$

and

$$f_K^+(t) - t\left(\frac{\bar{\tau}}{\tau}\right) f_K^-(t) = (1-t)\left(\frac{\bar{\tau}}{\tau}\right)^{1/2} f_K^0(t) \tag{2.8}$$

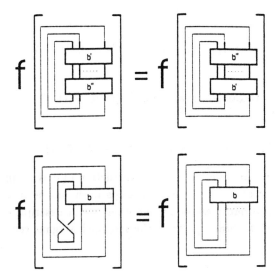

Fig. 2.3. Geometric representation of Eqs.(2.2).

where $f_K^+ \equiv f\{b' \sigma_i b''\}$; $f_K^- \equiv f\{b' \sigma_i^{-1} b''\}$; $f_K^0 \equiv f\{b' b''\}$ and the fraction $\frac{\bar{\tau}}{\tau}$ depends on the used representation.

3. The tensor representations of the braid generators can be written as follows

$$\sigma_i(u) = \lim_{u \to \infty} \sum_{klmn} R_{ln}^{km}(u) I^{(1)} \otimes \cdots I^{(i-1)} \otimes E_{nk}^i \otimes E_{ml}^{i+1} \otimes I^{(i+1)} \otimes \cdots I^{(n)}$$

$$(2.9)$$

where $I^{(i)}$ is the identity matrix acting in the position i; E_{nk} is a matrix with $(E_{nk})_{pq} = \delta_{np}\delta_{kq}$ and R_{ln}^{km} is the matrix satisfying the Yang-Baxter equation

$$\sum_{abc} R_{cr}^{bq}(v) R_{kc}^{ap}(u+v) R_{jb}^{ia}(u) = \sum_{abc} R_{bq}^{ap}(u) R_{cr}^{ia}(u+v) R_{ka}^{jb}(v) \qquad (2.10)$$

In that scheme both known polynomial invariants (Jones and Alexander) ought to be considered. In particular, it has been discovered in [18,19] that the solutions of Eq.(2.10) associated with the groups $SU_q(2)$ and $GL(1,1)$ are linked to Jones and Alexander invariants correspondingly. To be more specific:

a) $\frac{\bar{\tau}}{\tau} = t^2$ for Jones invariants, $f_K(t) \equiv V(t)$. The corresponding skein relations are

$$t^{-1}V^+(t) - tV^-(t) = (t^{-1/2} - t^{1/2})V^0(t) \qquad (2.11)$$

and
b) $\dfrac{\bar{\tau}}{\tau} = t^{-1}$ for Alexander invariants, $f_K(t) \equiv \nabla(t)$. The corresponding skein relations[¶] are

$$\nabla^+(t) - \nabla^-(t) = (t^{-1/2} - t^{1/2})\nabla^0(t) \qquad (2.12)$$

To complete this brief review of construction of polynomial invariants from the representation of the braid groups it should be mentioned that the Alexander invariants allow also another useful description [21]. Write the generators of the braid group in the so-called Magnus representation

$$\sigma_j \equiv \hat{\sigma}_j = \begin{pmatrix} 1 & 0 & \cdots & & & \\ 0 & \ddots & & & & \\ \vdots & & \boxed{A} & & \vdots & \\ & & & \ddots & 0 & \\ \cdots & & & 0 & 1 \end{pmatrix} \leftarrow j\text{th row}; \quad A = \begin{pmatrix} 1 & 0 & 0 \\ t & -t & 1 \\ 0 & 0 & 1 \end{pmatrix}$$

$$(2.13)$$

Now the Alexander polynomial of the knot represented by the closed braid $W = \prod_{j=1}^{N} \sigma_{\alpha_j}$ of the length N one can write as follows

$$(1 + t + t^2 + \ldots + t^{n-1})\, \nabla(t)\{A\} = \det\left[\prod_{j=1}^{N} \hat{\sigma}_{\alpha_j} - e\right] \qquad (2.14)$$

where index j runs "along the braid", i.e. labels the number of used generators, while the index $\alpha = \{1, \ldots, n-1, n, \ldots, 2n-2\}$ marks the set of braid generators (letters) ordered as follows $\{\sigma_1, \ldots, \sigma_{n-1}, \sigma_1^{-1}, \ldots, \sigma_{n-1}^{-1}\}$. In our further investigations we repeatedly address to that representation.

It should be emphasized that in general the length of the minimal irreducible length of the braid, introduced above, is not directly related to any topological knot invariants. Nevertheless the "primitive word" can serve as a well defined characteristic of the "knot complexity". The "primitive word" has the simple topological sense which can be expressed in the

---

[¶]Let us stress that the standard skein relations for Alexander polynomials one can obtain from Eq.(2.12) replacing $t^{1/2}$ by $-t^{1/2}$.

following necessary condition: if the "primitive word" of some closed braid of n strings has the unit length then this braid belongs to the "trivial" class and the corresponding knot is represented by the set of n disjoint unentangled trivial loops uniquely.

We are interested in the limit behavior of the knot or link invariants when the length of the corresponding braid tends to infinity, i.e. when the braid "grows". In this case we can rigorously define some topological characteristics, simpler than the algebraic invariant, which we call *the knot complexity*.

Definition 2 *Call the* **knot complexity,** η, *the power of some algebraic invariant,* $f_K(t)$ (Alexander, Jones, HOMFLY) (see also [22])

$$\eta = \lim_{|t| \to \infty} \frac{\ln f_K(t)}{\ln |t|} \qquad (2.15)$$

Remark. By definition, the "knot complexity" takes one and the same value for rather broad class of topologically different knots corresponding to algebraic invariants of one and the same power, being from this point of view weaker topological characteristics than complete algebraic polynomial.

The polynomial invariant can give exhaustive information about the knot topology. However when dealing with statistics of randomly generated knots, we frequently look for rougher characteristics of "topologically similar" knots. A similar problem arises in statistical mechanics when passing from the microcanonical ensemble to the Hibbs one: we lose some information about details of particular realization of the system but acquire smoothness of the measure and are able to apply standard thermodynamic methods to the system in question.

Let us summarize the advantages of the quantity η introduced in Eq.(2.15) with respect to the corresponding topological invariant $f_K(t)$:

(i) One and the same value of η characterizes a narrow class of "topologically similar" knots which is, however, much broader than the class represented by the polynomial invariant $X(t)$. This enables us to introduce the smoothed measures and distribution functions for η.

(ii) The knot complexity η describes correctly (at least from the physical point of view) the limit cases: $\eta = 0$ corresponds to "weakly entangled" trajectories whereas $\eta \sim N$ matches the system of "strongly entangled" paths. The latter case has been discussed at length in [22].

(iii) The knot complexity keeps all nonabelian properties of the polynomial invariants.

The main purpose of the present section is the estimation of the limit probability distribution of η for the knots obtained by randomly generated closed B_3-braids of the length N. It should be emphasized that we essentially simplify the general problem "of knot entropy". Namely, we introduce an additional requirement that the knot should be represented by a braid from the group B_3 without fail.

STATISTICAL MODEL. We begin the investigation of the probability properties of algebraic knot invariants by analyzing statistics of the random loops ("brownian bridges") on simplest noncommutative groups. Most generally the problem can be formulated as follows.

Take the discrete group $\mathcal{G}_n$ with a fixed finite number of generators $\{g_1, \ldots, g_{n-1}\}$. Let ν be the uniform distribution on the set $\{g_1, \ldots, g_{n-1}, g_1^{-1}, \ldots, g_{n-1}^{-1}\}$. For convenience we suppose $h_j = g_i$ for $j = i$ and $h_j = g_i^{-1}$ for $j = i+n-1$; $\nu(h_j) = \frac{1}{2n-2}$ for any j. We construct the (right-hand) side random walk (the random word) on $\mathcal{G}_n$ with a transition measure ν, i.e. the Markov chain $\{\xi_n\}$, $\xi_0 = e \in \mathcal{G}_n$ and $\mathrm{Prob}(\xi_j = u | \xi_{j-1} = v) = \nu(v^{-1}u) = \frac{1}{2n-2}$. It means that with the probability $\frac{1}{2n-2}$ we add the element h_{α_N} to the given word $h_{N-1} = h_{\alpha_1} h_{\alpha_2} \ldots h_{\alpha_{N-1}}$ from the right-hand side $^{\|}$.

Definition 3 *The random word W formed by N letters taken independently with the uniform probability distribution $\nu = \frac{1}{2n-2}$ from the set $\{g_1, \ldots, g_{n-1}, g_1^{-1}, \ldots, g_{n-1}^{-1}\}$ is called the* **brownian bridge** *(BB) of length N on the group $\mathcal{G}_n$ if the primitive word of W is identical to the unity.*

Two questions require most of our attention:

1. What is the probability distribution $P(N)$ of the brownian bridge on the group $\mathcal{G}_n$.

2. What is the conditional probability distribution $P(k, m|N)$ of the fact that the subword W' consisting of first m letters of the N-letter word W has the primitive path k under the condition that the whole word W is the brownian bridge on the group $\mathcal{G}_n$. (Hereafter $P(k, m|N)$ is referred to as the conditional distribution for BB.)

$^{\|}$Analogously we can construct the left-hand side Markov chain.

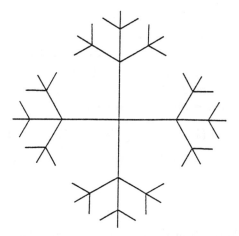

Fig. 2.4. Cayley tree $C(\Gamma_2)$ corresponding to the free group Γ_2.

It has been shown in the paper [9] that for the free group the corresponding problem can be mapped on the investigation of the random walks on the simply connected tree. Below we represent shortly some results concerning the limit behavior of the conditional probability distribution of BB on the Cayley tree. In the case of braids the more complicated group structure does not allow us to apply the same simple geometrical image directly. Nevertheless the problem of the limit distribution for the random walks on B_n can be reduced to the consideration of the random walk on some graph $C(\Gamma)$. In case of the group B_3 we are able to construct this graph evidently, whereas for the group B_n $(n \geq 4)$ we give upper estimations for the limit distribution of the random walks considering the statistics of Markov chains on so-called local groups.

2.2.2. *Statistics of Random Walks and Joint Distributions of Brownian Bridges on Free Group*

The free group, Γ_2, with two generators g_1 and g_2 has a well-known matrix representation (see, for instance, [23])

$$g_1 = \begin{pmatrix} 1 & 0 \\ 2 & 1 \end{pmatrix}; \qquad g_2 = \begin{pmatrix} 1 & 2 \\ 0 & 1 \end{pmatrix} \qquad (2.16)$$

Consider the Markov chain with the states in the set $\{g_1, g_2, g_1^{-1}, g_2^{-1}\}$ as it is described in the previous section. Due to the simple topological structure of the free group, the limit distribution of the random walk on the group Γ_2 is mapped on the limit distribution of the random paths on the Cayley $C(\Gamma_2)$ tree with 4 branches [3,11,9] and the local transitional probabilities equal to $\frac{1}{4}$ (see fig.2.4). In particular, the probability, $P(k, N)$, of the fact that in the randomly generated N-letter word W the primitive word length is k, satisfies the set of equations [14]

$$P(k, N+1) = \tfrac{1}{4}P(k+1, N) + \tfrac{3}{4}P(k-1, N); \qquad (k \geq 2)$$

$$P(k, N+1) = \tfrac{1}{4}P(k+1, N) + P(k, N); \qquad (k = 1)$$

$$P(k, N+1) = \tfrac{1}{4}P(k+1, N); \qquad (k = 0)$$

$$P(k, 0) = \delta_{k,0}$$

$$(2.17)$$

The solution of Eq.(2.17) in the limit $N \to \infty$ near the maximum of the distribution function is:

$$P(k, N) = \left(\frac{3}{4}\right)^{N/2} 3^{(k+1)/2} \, Q(k+1, N) \qquad (2.18)$$

where

$$Q(k, N) \simeq \begin{cases} 3\sqrt{\dfrac{2}{\pi}} \dfrac{1}{N^{3/2}} & (k = 1) \\[4mm] 2\sqrt{\dfrac{2}{\pi}} \dfrac{k}{N^{3/2}} \exp\left\{-\dfrac{k^2}{2N}\right\} & (1 \ll k < N) \end{cases} \qquad (2.19)$$

The function $Q(k, N)$ defines the probability distribution for the simplest random walk on the half-line $\mathbb{Z}^+$ with the boundary condition $Q_W(k = 0, N) = 0$.

For exact solution of Eq.(2.17) we can address to the reference [15].

Lemma 1 *The limit conditional probability distribution, $P(k, m|N)$, for the brownian bridge on the group Γ_2 obeys the central limit theorem* [9]

$$P(k, m|N) = \sqrt{\frac{2}{\pi}} \frac{k^2}{(m(N-m))^{3/2}} \exp\left\{-\frac{k^2}{2}\left(\frac{1}{m} + \frac{1}{N-m}\right)\right\} \qquad (2.20)$$

when $N \to \infty$; $\frac{m}{N} = const$ *and* $1 \ll k < N$.

Proof. According to the definition of conditional probability distribution on BB, we split the whole word W in two subwords W' and W'' having m and $(N - m)$ letters correspondingly. Now using the Definition

3 and the fact that the word W is realized as a Markov chain, we can represent the conditional distribution function $P(k, m|N)$ in the following form

$$P(k, m|N) = \frac{P(k, m)\, P(k, N - m)}{P(0, N)\, \mathcal{N}(k)} \qquad (2.21)$$

where $\mathcal{N}(k) = 4 \cdot 3^{k-1}$ is the number of different primitive words of the length k.

To make the Eq.(2.21) clearer recall that the N-letter word W on the group Γ_2 is in one-to-one correspondence with the N-step trajectory on the Cayley tree $C(\Gamma_2)$ and the length of the primitive word W is identical to the distance between the ends of the given trajectory along the tree (i.e. is equal to the geodesics). The functions $P(k, m)$ and $P(k, N - m)$ give the probability of the fact that the $m-$ and $(N - m)$–step paths have ended in an *arbitrary* points of the Cayley tree on the distance k from the origin. The probability of coincidence of the ends of these two different paths in some *common* point on the distance k from the origin is $\frac{1}{\mathcal{N}(k)}$.

Substituting Eqs.(2.18), (2.19) into Eq.(2.21) we obtain the postulated expression Eq.(2.20), where the pre-exponent is due to the Dirichlet boundary condition at $k = 0$ $\square$.

Lemma 2 *The joint conditional distribution $P(k_1, m_1; \ldots; k_s, m_s|N)$ of the BB on the group Γ_2 is converged for $N \to \infty$ (where $\sum_{j=1}^{s} m_j = N$; $\frac{m_j}{N} = \text{const}$ and $1 \ll k_j \ll N$ for any $1 < j < s$) to the finite-dimensional distribution of the BB on the halfline $\mathbb{Z}^+$.*

Proof. Let us define the two-point conditional distribution functions, $\pi^+(k_1, m_1; k_2^+, m_2|N)$ and $\pi^-(k_1, m_1; k_2^-, m_2|N)$, in the sense of the probabilities of two following events satisfied simultaneously:

i) The first m_1-letter subword W' in the N-letter word W has the primitive length k_1;

ii) The subword W'' in the same N-letter word obtained by adding the next letter to the subword W' ($m_2 = m_1 + 1$) has the primitive length $k_2^+ = k_1 + 1$ (for π^+) or $k_2^- = k_1 - 1$ (for π^-) under the condition that the whole word W is fully contractible (i.e. its primitive length is equal to zero).

Obviously, $\pi^{\pm}(k_1, m_1; k_2, m_2|N)$ give the local transitional probabilities for the conditional random walk when we make one step "forth" or "back" along the geodesics on the Cayley tree (fig.2.5). Now in order

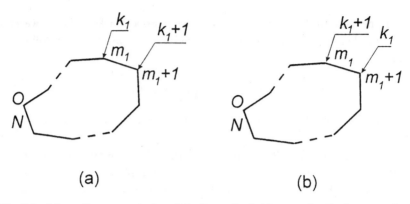

Fig. 2.5. Schematic representation of the brownian bridges on the Cayley tree. Cases (a) and (b) correspond to calculation of π^+ and π^-.

to prove that the conditional radial ** random process on the group Γ_2 is mapped on the simplest random walk without any drift onto $\mathbb{Z}^+$ and has the Wiener messier, it is sufficient to show that $\pi^+ = \pi^- = \frac{1}{2}$ when $N \to \infty$; i.e. the condition of the contractibility of the whole N-step trajectory eliminates the drift from the origin on the Cayley tree for the local jumps.

1. Suppose $k_2^+ = k_1 + 1$. Under the condition (ii) we have

$$\pi^+(k_1, m_1; k_2^+ = k_1 + 1, m_2 = m_1 + 1|N) =$$

$$\frac{P(k_1, m_1)\, P^+(k_2^+ - k_1, 1)\, P(k_2^+, N - m_1 - 1)}{P(0, N)\, \mathcal{N}(k_1)\, (z - 1)} \qquad (2.22)$$

where: $(z - 1)$ is the number of the tree branches connecting one arbitrary point on the tree to the points on the next coordinational sphere (z is the coordinational number of the Cayley tree), $z = 4$; $P^+(k_2^+ - k_1, 1)$ is the probability to increase the distance along the tree per one unit making one random step for $k_1 \geq 1$, $P^+ = \frac{z-1}{z} = \frac{3}{4}$.

Substituting Eq.(2.18) into Eq.(2.22) we obtain the following expression for π^+

$$\pi^+ = \frac{3\sqrt{3}}{8}\, \frac{Q(k_1 + 1, m_1)\, Q(k_1 + 2, N - m_1 - 1)}{Q(1, N)} \qquad (2.23)$$

** The distances are measured in terms of lengths of geodesics on the Cayley tree.

2. Now let $k_2^- = k_1 - 1$. Reversing the direction along the trajectory, we get

$$
\begin{aligned}
\pi^- (k_1, m_1; k_2^- = k_1 - 1, m_2 = m_1 + 1 | N) &\equiv \\
\pi^+ (k_2^-, N - m_1 - 1; k_2^- + 1, m_1 | 0) &= \\
\frac{P(k_2^-, N - m_1 - 1) \, P^+(k_1 - k_2^-, 1) \, P(k_2^- + 1, m_1)}{P(0, N) \, \mathcal{N}(k_1) \, (z - 1)}
\end{aligned}
\tag{2.24}
$$

where $P^+(k_1 - k_2^-, 1) = \frac{3}{4}$ (compare to Eq.(2.22)).

Eq.(2.24) signifies the fact that the probability does not change if the random word is written in reversed order of steps, i.e., the 1st step has number N, the 2nd has number $(N - 1)$ and so on. Thus, π^- has the form similar to Eq.(2.22) and it can be written as

$$
\pi^- = \frac{3\sqrt{3}}{8} \frac{Q(k_1 + 1, m_1) \, Q(k_1, N - m_1 - 1)}{Q(1, N)}
\tag{2.25}
$$

Using the probability conservation law

$$
\pi^+ + \pi^- = 1
$$

and the recursion relation for the simplest random walk on the half-line $\mathbb{Z}^+$ (extracted from Eqs.(2.17)-(2.18))

$$
Q(k_1 + 2, N - m_1 - 1) + Q(k_1, N - m_1 - 1) = 2Q(k_1 + 1, N - m_1), \quad (k \geq 1)
$$

it is possible to rewrite $\pi^\pm$ as follows:

$$
\begin{aligned}
\pi^+ &= \frac{\pi^+}{\pi^+ + \pi^-} = \frac{1}{2} \frac{Q(k_1 + 2, N - m_1 - 1)}{Q(k_1 + 1, N - m_1)} \\
\pi^- &= \frac{\pi^-}{\pi^+ + \pi^-} = \frac{1}{2} \frac{Q(k_1, N - m_1 - 1)}{Q(k_1 + 1, N - m_1)}
\end{aligned}
\tag{2.26}
$$

Substituting Eq.(2.19) into Eq.(2.26) we find

$$
\pi^+ = \frac{1}{2} - \frac{c(1 - s)}{\sqrt{N}}; \qquad \pi^- = \frac{1}{2} + \frac{c(1 - s)}{\sqrt{N}}
\tag{2.27}
$$

where $c = k_1/\sqrt{N}$ $(1 \ll k_1 \ll N)$, $s = m_1/N$ $(1 < m < N)$ and $N \to \infty$.

Thus, the transition probabilities for local jumps along the geodesics on the Cayley tree under the condition of BB coincide with the transition probabilities for the simplest random walk on the halfline $\mathbb{Z}^+$ when $N \to \infty$. Hence, we have one-to-one mapping of the "radial" random walk on the tree under the condition of BB onto the standard diffusion process *without the drift* on the halfline. Applying the standard central limit theorem to the last process we get the desired statement of the Lemma 2 □.

2.2.3. *Random Walks on* $PSL(2,\mathbb{Z})$, B_3 *and Distribution of Powers of Alexander Invariants*

We begin our analysis with the calculation of the distribution function for the conditional BB on the simplest nontrivial braid group B_3. The group B_3 can be represented by 2×2 matrices. To be specific, the braid generators σ_1 and σ_2 in the Magnus representation [21] look as follows:

$$\sigma_1 = \begin{pmatrix} -t & 1 \\ 0 & 1 \end{pmatrix}; \qquad \sigma_2 = \begin{pmatrix} 1 & 0 \\ t & -t \end{pmatrix}, \tag{2.28}$$

where t is "the spectral parameter". It is well known that for $t = -1$ the matrices σ_1 and σ_2 generate the group $PSL(2,\mathbb{Z})$ in such a way that the whole group B_3 is its central extension with the centre

$$(\sigma_1\sigma_2\sigma_1)^{4\lambda} = (\sigma_2\sigma_1\sigma_2)^{4\lambda} = (\sigma_1\sigma_2)^{6\lambda} = (\sigma_2\sigma_1)^{6\lambda} = \begin{pmatrix} t^{6\lambda} & 0 \\ 0 & t^{6\lambda} \end{pmatrix} \tag{2.29}$$

First restrict ourselves with the examination of the group $PSL(2,\mathbb{Z})$, for which we define $\tilde{\sigma}_1 = \sigma_1$ and $\tilde{\sigma}_2 = \sigma_2$ (at $t = -1$).

The canonical representation of $PSL(2,\mathbb{Z})$ is given by the unimodular matrices S, T:

$$S = \begin{pmatrix} 0 & 1 \\ -1 & 0 \end{pmatrix}; \qquad T = \begin{pmatrix} 1 & 1 \\ 0 & 1 \end{pmatrix} \tag{2.30}$$

The braiding relation $\tilde{\sigma}_1\tilde{\sigma}_2\tilde{\sigma}_1 = \tilde{\sigma}_2\tilde{\sigma}_1\tilde{\sigma}_2$ in the $\{S, T\}$-representation takes the form

$$S^2TS^{-2}T^{-1} = 1 \tag{2.31}$$

in addition we have

$$S^4 = (ST)^3 = 1 \tag{2.32}$$

This representation is well known and signifies the fact that in terms of $\{S, T\}$-generators the group $SL(2,\mathbb{Z})$ is a free product $Z^2 \otimes Z^3$ of two cyclic groups of the 2nd and the 3rd orders correspondingly.

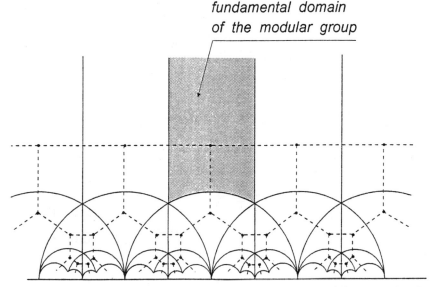

Fig. 2.6. The Riemann surface for the modular group The graph $C(\Gamma)$ representing the topological structure of $PSL(2, \mathbb{Z})$ is shown by the dashed line.

The connection of $\{S, T\}$ and $\{\tilde{\sigma}_1, \tilde{\sigma}_2\}$ is as follows

$$
\begin{aligned}
\tilde{\sigma}_1 &= T && (T = \tilde{\sigma}_1) \\
\tilde{\sigma}_2 &= T^{-1}ST^{-1} && (S = \tilde{\sigma}_1\tilde{\sigma}_2\tilde{\sigma}_1)
\end{aligned}
\tag{2.33}
$$

RANDOM WALKS ON GROUP $PSL(2,\mathbb{Z})$. The modular group $PSL(2,\mathbb{Z})$ is a discrete subgroup of the group $PSL(2,\mathbb{R})$. The fundamental domain of $PSL(2,\mathbb{Z})$ has the form of a circular triangle ABC with angles $\left\{0, \frac{\pi}{3}, \frac{\pi}{3}\right\}$ situated in the upper halfplane $\text{Im}\zeta > 0$ of the complex plane $\zeta = \xi + i\eta$ (see fig.2.6 for details). According to the definition of the fundamental domain, at least one element of each orbit of $PSL(2,\mathbb{Z})$ lies inside ABC-domain and two elements lie on the same orbit if and only if they belong to the boundary of the ABC-domain. The group $PSL(2,\mathbb{Z})$ is completely defined by its basic substitutions under the action of generators S and T:

$$
\begin{aligned}
S &: \quad \zeta \to -1/\zeta \\
T &: \quad \zeta \to \zeta + 1
\end{aligned}
\tag{2.34}
$$

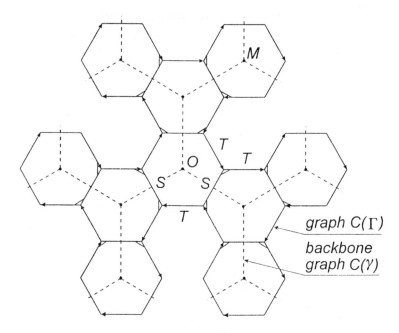

Fig. 2.7. The graph $C(\Gamma)$ and its backbone graph $C(\gamma)$ (see the explanations in the text).

Let us choose an arbitrary element ζ_0 from the fundamental domain and construct a corresponding orbit. In other words, we raise a graph, $C(\Gamma)$, which connects the neighboring images of the initial element ζ_0 obtained under successive action of the generators from the set $\{S, T, S^{-1}, T^{-1}\}$ to the element ζ_0. The corresponding graph is shown in the fig.2.6 by the broken line and its topological structure is clearly reproduced in fig.2.7. It can be seen that although the graph $C(\Gamma)$ does not correspond to the free group and has local cycles, its "backbone", $C(\gamma)$, has Cayley tree structure but with the reduced number of branches as compared to the free group $C(\Gamma_2)$.

Let us turn to the problem of limit distribution of a random walk on the graph $C(\Gamma)$. The walk is determined as follows:

1. Take an initial point ("root") of the random walk on the graph $C(\Gamma)$. Consider the discrete random jumps over the neighboring vertices of the graph with the transition probabilities induced by the uni-

form distribution ν on the set of generators $\{\tilde{\sigma}_1, \tilde{\sigma}_2, \tilde{\sigma}_1^{-1}, \tilde{\sigma}_2^{-1}\}$. These probabilities are (see Eq.(2.33))

$$\text{Prob}(\xi_n = T\zeta_0 \mid \xi_{n-1} = \zeta_0) = \tfrac{1}{4}$$
$$\text{Prob}(\xi_n = (T^{-1}ST^{-1})\zeta_0 \mid \xi_{n-1} = \zeta_0) = \tfrac{1}{4}$$
$$\text{Prob}(\xi_n = T^{-1}\zeta_0 \mid \xi_{n-1} = \zeta_0) = \tfrac{1}{4} \qquad (2.35)$$
$$\text{Prob}(\xi_n = (TS^{-1}T)\zeta_0 \mid \xi_{n-1} = \zeta_0) = \tfrac{1}{4}$$

The following facts should be taken into account:

(a) the elements $S\zeta_0$ and $S^{-1}\zeta_0$ represent one and the same point, i.e. coincide (as it follows from Eq.(2.34));

(b) the process is marcovian in terms of the alphabet $\{\tilde{\sigma}_1, \ldots, \tilde{\sigma}_2^{-1}\}$ *only*;

(c) the total transition probability is conserved.

2. Define the shortest distance, k, along the graph between the root and terminal points of the random walk. According to its construction, this distance coincides with the length $|W_{\{S,T\}}|$ of the minimal irreducible word $W_{\{S,T\}}$ written in the alphabet $\{S, T, S^{-1}, T^{-1}\}$. The link of the distance, k, with the length $|W_{\{\tilde{\sigma}_1, \tilde{\sigma}_2\}}|$ of the minimal irreducible word $W_{\{\tilde{\sigma}_1, \tilde{\sigma}_2\}}$ written in terms of the alphabet $\{\tilde{\sigma}_1, \tilde{\sigma}_2, \tilde{\sigma}_1^{-1}, \tilde{\sigma}_2^{-1}\}$ is established in the following lemma.

Lemma 3 (a) $|W_{\{\tilde{\sigma}_1, \tilde{\sigma}_2\}}| = 0$ *if and only if* $k = 0$; (b) *for* $k \gg 1$ *the length* $|W_{\{\tilde{\sigma}_1, \tilde{\sigma}_2\}}|$ *has the following behavior*

$$\left. \frac{|W_{\{\tilde{\sigma}_1, \tilde{\sigma}_2\}}|}{k} \right|_{k \to \infty} = 1 + O\left(\frac{1}{k}\right)$$

The proof, rather trivial, is based on the evident construction of the graph $C(\Gamma)$ where each bond can be associated with the generators $\tilde{\sigma}_1^{\pm 1}$ and $(\tilde{\sigma}_1 \tilde{\sigma}_2 \tilde{\sigma}_1)^{\pm 1}$ by means of Eqs.(2.33).

We define the "coordinates" of the graph vertices in the following way (see fig.2.7):

(a) We apply the arrows to the bonds of the graph Γ corresponding to T-generators. The step towards (backwards) the arrow means the application of T (T^{-1}).

(b) We characterize each elementary cell of the graph Γ by its distance, μ, along the graph backbone γ from the root cell.

(c) We introduce the variable $\alpha = \{1,2\}$ which numerates the vertices *in each cell* only. We assume that the walker stays in the cell M located at the distance μ along the backbone from the origin if and only if it visits one of two ingoing vertices of M. Such labelling gives unique coding of the whole graph $C(\Gamma)$.

Define the probability $U_\alpha(\mu, N)$ of the fact that the N-step random walk along the graph $C(\Gamma)$ starting from the root point is ends in α-vertex of the cell on the distance of μ steps along the backbone. It should be emphasized that $U_\alpha(\mu, N)$ is the probability to stay in *any* of $\mathcal{N}_\gamma(\mu) = 3 \cdot 2^{\mu-1}$ cells situated at the distance μ along the backbone.

It is possible to write the closed system of recursion relations for the functions $U_\alpha(\mu, N)$. However, here we attend to rougher characteristics of random walk. Namely, we calculate the "integral" probability distribution of the fact that the trajectory of the random walk starting from an arbitrary vertex of the root cell O has ended in an arbitrary vertex point of the cell M situated on the distance μ along the graph backbone. This probability, $U(\mu, N)$, reads

$$U(\mu, N) = \frac{1}{2} \sum_{\alpha=\{1,2\}} U_\alpha(\mu, N)$$

Lemma 4 *The relation between the distances k, along the graph Γ, and μ along its backbone γ is as follows:*

$$\left. \frac{k}{\mu} \right|_{\mu \to \infty} = 1 + O\left(\frac{1}{\mu}\right) \tag{2.36}$$

This fact is the result of the constructions of the graphs $C(\Gamma)$ and $C(\gamma)$ (fig.2.7).

The following theorem gives the limit distribution for the random walks on the group $PSL(2,\mathbb{Z})$.

Theorem 5 *The probability distribution $U(k, N)$ of the fact that the randomly generated N-letter word $W_{\{\tilde{\sigma}_1, \tilde{\sigma}_2\}}$ with the uniform distribution $\nu = \frac{1}{4}$ over the generators $\{\tilde{\sigma}_1, \tilde{\sigma}_2, \tilde{\sigma}_1^{-1}, \tilde{\sigma}_2^{-1}\}$ can be contracted to the min-*

imal irreducible word of length k, has the following limit behavior

$$U(k, N) \simeq \frac{h}{\sqrt{\pi(4-h)}} \left(\frac{h}{4(h-2)}\right)^N \begin{cases} \dfrac{1}{N^{3/2}} & k = 0 \\[2mm] \dfrac{k}{N^{3/2}} 2^{k/2} \exp\left(-\dfrac{k^2 h}{4N}\right) & 1 \ll k \end{cases}$$

$$(2.37)$$

where $h = 2 + \dfrac{\sqrt{2}}{2}$.

Proof. Suppose the walker stays in the vertex α of the cell M located at the distance $\mu > 1$ from the origin along the graph backbone $C(\gamma)$. The change in μ after making of one arbitrary step from the set $\{\tilde{\sigma}_1, \tilde{\sigma}_2, \tilde{\sigma}_1^{-1}, \tilde{\sigma}_2^{-1}\}$ is summarized in the following table:

$\alpha = 1$		$\alpha = 2$	
$\tilde{\sigma}_1 = T$	$\mu \to \mu + 1$	$\tilde{\sigma}_1 = T$	$\mu \to \mu - 1$
$\tilde{\sigma}_2 = T^{-1}ST^{-1}$	$\mu \to \mu$	$\tilde{\sigma}_2 = T^{-1}ST^{-1}$	$\mu \to \mu + 1$
$\tilde{\sigma}_1^{-1} = T^{-1}$	$\mu \to \mu - 1$	$\tilde{\sigma}_1^{-1} = T^{-1}$	$\mu \to \mu + 1$
$\tilde{\sigma}_2^{-1} = TS^{-1}T$	$\mu \to \mu + 1$	$\tilde{\sigma}_2^{-1} = TS^{-1}T$	$\mu \to \mu$

It is clear that for any value of α two steps increase the length of the backbone, μ, one step decreases it and one step leaves μ without changes.

Let us introduce the effective probabilities: p_1 – to jump to some specific cell among 3 neighboring ones of the graph $C(\Gamma)$ and p_2 – to stay in the given cell. Because of the symmetry of the graph, the conservation law has to be written as $3p_1 + p_2 = 1$. By definition we have: $p_1 \overset{def}{=} \nu = \frac{1}{4}$. Thus we can write the following set of recursion relations for the integral probability $U(\mu, N)$

$$U(\mu, N+1) = \tfrac{1}{4}U(\mu+1, N) + \tfrac{1}{4}U(\mu, N) + \tfrac{1}{2}U(\mu-1, N) \quad (\mu \geq 2)$$

$$U(\mu, N+1) = \tfrac{1}{4}U(\mu+1, N) + \tfrac{1}{2}U(\mu, N) \quad\quad\quad\quad (\mu = 1)$$

$$U(\mu, N = 0) = \delta_{\mu,1}$$

$$(2.38)$$

The solution of Eq.(2.38) we search in the form

$$U(\mu, N) = A^\mu B^N V(\mu, N) \tag{2.39}$$

where the constants A and B are chosen from the auxiliary conditions:

$$\frac{A}{4B} = \frac{1}{h}; \qquad \frac{1}{4B} = 1 - \frac{2}{h}; \qquad \frac{1}{2AB} = \frac{1}{h} \qquad (h > 1) \qquad (2.40)$$

Resolving these equations we get:

$$A = \sqrt{2}; \quad B = \frac{1}{4} + \frac{\sqrt{2}}{2}; \quad h = 2 + \frac{\sqrt{2}}{2} \qquad (2.41)$$

The equations (2.40) imply that for the function $V(\mu, N)$ we obtain a usual 1D random walk on the halfline $\mu \geq 0$ (i.e. $V(\mu \leq 0, N) \equiv 0$) with conserved transition probabilities and with some special boundary and initial conditions:

$$V(\mu, N+1) = \tfrac{1}{h}V(\mu+1, N) + \left(1 - \tfrac{2}{h}\right)V(\mu, N) + \tfrac{1}{h}V(\mu-1, N) \quad (\mu \geq 2)$$
$$V(\mu, N+1) = \tfrac{1}{h}V(\mu+1, N) + 2\left(1 - \tfrac{2}{h}\right)V(\mu, N) \qquad (\mu = 1)$$
$$V(\mu, N = 0) = \delta_{\mu,1}$$

$$(2.42)$$

It is practicable to obtain the exact asymptotic solution of Eq.(2.42) for $N \to \infty$. We begin by rewriting Eqs.(2.42) as

$$V(\mu, N+1) = \frac{1}{h}V(\mu+1, N) + \left(1 - \frac{2}{h}\right)(1 + \delta_{\mu,1})V(\mu, N) + \frac{1}{h}V(\mu-1, N)$$

$$(2.43)$$

with the boundary $V(\mu = 0, N) = 0$ and initial $V(\mu, N = 0) = \delta_{\mu,1}$ conditions.

Then we introduce the generating function for N-variable and the sin-Fourier transform for μ-variable on the halfline $\mu \geq 0$

$$V(u, s) = \sum_{N=0}^{\infty} s^N \sum_{\mu=0}^{\infty} V(\mu, N) \sin\frac{\pi u \mu}{l} \qquad (2.44)$$

From Eqs.(2.43)-(2.44) we have

$$\frac{1}{s}V(u, s) - \frac{1}{s}\sin\frac{\pi u}{l} = \frac{2}{h}\cos\frac{\pi u}{l}V(u, s) + \left(1 - \frac{2}{h}\right)V(u, s)$$
$$+ \left(1 - \frac{1}{h}\right)\sin\frac{\pi u}{l}\frac{1}{l}\int_0^l \sin\frac{\pi u}{l}V(u, s)du$$

$$(2.45)$$

The solution of Eq.(2.45) reads

$$\frac{1}{l}\int_0^l \sin\frac{\pi u}{l}V(u,s)du = \frac{D_1(h,s)}{D_2(h,s)} \tag{2.46}$$

where

$$D_1(h,s) = \frac{1}{\pi}\int_0^\pi \frac{\sin^2 w\,dw}{1 - s\left(\frac{2}{h}\cos w + 1 - \frac{2}{h}\right)}$$

$$= -\frac{h}{s} + \frac{h}{4s^2}\left(h + 6s - hs - \sqrt{h}\sqrt{(1-s)(h+4s-hs)}\right)\Big|_{s\to 1^-} \tag{2.47}$$

$$\simeq \frac{h}{2} - \frac{h\sqrt{h}}{2}\sqrt{1-s} + O(1-s)$$

and

$$D_2(h,s) = 1 - \left(1 - \frac{2}{h}\right)\frac{1}{\pi}\int_0^\pi \frac{s\sin^2 w\,dw}{1 - s\left(\frac{2}{h}\cos w + 1 - \frac{2}{h}\right)}$$

$$= 1 - \left(1 - \frac{2}{h}\right)sD_1(h,s) \tag{2.48}$$

It is easy to see that $D_2(h,s)$ is always positive for any $|s| \leq 1$. It means that equation (2.45) has a continuous spectrum and the limit distribution of the function $V(\mu, N)$ is governed by the central limit theorem for the random walks on the halfline. The exact solution for the function $V(\mu, s)$ is

$$V(\mu, s) = \frac{1}{D_2(h,s)}\frac{2}{\pi}\int_0^\pi \frac{\sin w \sin w\mu\,dw}{1 - s\left(\frac{2}{h}\cos w + 1 - \frac{2}{h}\right)} \tag{2.49}$$

In particular, we have

$$V(\mu = 1, s) \simeq \frac{2h\left(1 - \sqrt{h}\sqrt{1-s}\right)}{4 - h + (h-2)\sqrt{h}\sqrt{1-s}} = \frac{2\sqrt{h}}{h-2}\frac{1}{a+\sqrt{\epsilon}} + \text{const} \tag{2.50}$$

and

$$V(\mu \gg 1, s) \simeq \frac{2h\exp\left(-\mu\sqrt{h}\sqrt{1-s}\right)}{4 - h + (h-2)\sqrt{h}\sqrt{1-s}} = \frac{2\sqrt{h}}{h-2}\frac{\exp\left\{-\mu\sqrt{h}\sqrt{1-s}\right\}}{a+\sqrt{\epsilon}} \tag{2.51}$$

where $a = \frac{4-h}{\sqrt{h}(h-2)}$ and $\epsilon = 1 - s > 0$.

Performing the inverse Laplace transform and taking into account the contribution from the branching point at $\epsilon = 0$ only, we obtain in the limit $N \to \infty$

$$V(0, N) \simeq \frac{2\sqrt{h}}{h-2} \left(\frac{1}{\sqrt{\pi N}} - ae^{a^2 N} \mathrm{erfc}\left(a\sqrt{N}\right) \right)$$

$$\simeq \frac{\sqrt{h}}{a\sqrt{\pi}(h-2)} \frac{1}{N^{3/2}}$$

(2.52)

and

$$V(\mu \gg 1, N) \simeq \frac{2\sqrt{h}}{h-2} \left(\frac{1}{\sqrt{\pi N}} e^{-\frac{\mu^2 h}{4N}} - ae^{\mu a \sqrt{h} + a^2 N} \mathrm{erfc}\left(a\sqrt{N} + \frac{\mu\sqrt{h}}{2\sqrt{N}}\right) \right)$$

$$\simeq \frac{\sqrt{h}}{a\sqrt{\pi}(h-2)} \frac{\mu}{N^{3/2}} \exp\left(-\frac{\mu^2 h}{4N}\right)$$

(2.53)

Substituting the last equation in Eq.(2.39) and taking into account the Lemmas 3-4 we get the statement of the Theorem 5 $\square$.

Corollary 1 *The probability distribution $U(k, m|N)$ of the fact that in the randomly generated N-letter* **trivial** *word in the alphabet $\{\tilde{\sigma}_1, \tilde{\sigma}_2, \tilde{\sigma}_1^{-1}, \tilde{\sigma}_2^{-1}\}$ the subword of first m letters has a minimal irreducible length k reads*

$$U(k, m|N) = \frac{h}{\sqrt{\pi}(4-h)} \frac{k^2}{(m(N-m))^{3/2}} \exp\left\{ \frac{k^2 h}{4}\left(\frac{1}{m} + \frac{1}{N-m}\right) \right\}$$

(2.54)

Proof. The conditional probability distribution $U(\mu, m|N)$ of the fact that the random walk on the backbone graph, $C(\gamma)$, starting in the origin, visits after first m ($\frac{m}{N} = $ const) steps some graph vertex situated at the distance μ and after N steps returns to the origin, is determined as follows (compare to the proof of Lemma 1)

$$U(\mu, m|N) = \frac{U(\mu, m)U(\mu, N-m)}{U(\mu = 0, N)\mathcal{N}_\gamma(\mu)}$$

(2.55)

where $\mathcal{N}_\gamma = 3 \cdot 2^{\mu-1})$ and $U(\mu, N)$ is given by (2.37). Using Lemma 3 we get Eq.(2.54) $\square$.

STATISTICS OF BROWNIAN BRIDGES OF B_3 AND LIMIT BEHAVIOR OF POWERS OF ALEXANDER INVARIANTS. Now we are able to formulate some limit theorems for BB on the group B_3 as well as to find the limit distribution for the knot complexity η (i.e. power of the Alexander polynomial of the knots represented by the random braids from B_3).

Theorem 6 *The probability $Z(k, m|N)$ for the brownian bridge on the group B_3 has the limit behavior*

$$Z(k, m|N) \asymp \begin{cases} \dfrac{c_1}{m^{3/2}(N - m)^{3/2}} & k = 0 \\[4mm] \psi\left(\dfrac{k}{m}\right) \psi\left(\dfrac{k}{N - m}\right) \exp\left(-\dfrac{c_2\, Nk^2}{m(N - m)}\right) & k \gg 1 \end{cases}$$

(2.56)

where $\psi(k, m)$ is some function of k and m with power-low behavior in k and m. (We expect $\psi(k, m) \sim \frac{k}{m^{3/2}}$ but the given proof is too rough to show that behavior).

Proof. The conditional limit probability distribution $Z(k, m|N)$ for $N \to \infty$ is bounded from below and above

$$P(k, m|N) \leq Z(k, m|N) \leq U(k, m|N) \qquad (2.57)$$

where $P(k, m|N)$ and $U(k, m|N)$ are the limit probabilities for the brownian bridges on the groups Γ_2 (i.e. the free group) and $PSL(2, \mathbb{Z})$ (i.e. the braid group at the point $t = -1$) correspondingly. Substituting Eq.(2.57) for asymptotics Eq.(2.20) and Eq.(2.54), we come to the conclusion Eq.(2.56) $\square$.

The problem of calculating conditional limit probability distribution of the brownian bridges on the group B_3 can be easily turned into the problem of calculating conditional distribution function for the powers of Alexander polynomial invariants of knots produced by randomly generated closed braids from the group B_3 which allows one to make the first step in investigation of correlations in knotted random walks.

The closure of an arbitrary braid $b \in B_3$ of the total length N gives the knot (link) K. Split the braid b in two parts b' and b'' with the corresponding lengths m and $N - m$ and make the "phantom closure" of the subraids b' and b'' as it is shown in fig.2.8. The phantomly closed subraids b' and b'' correspond to the set of phantomly closed parts ("subknots") of the knot (link) K. The next question is what the conditional probability to find these subknots in the state characterized by the complexity η when

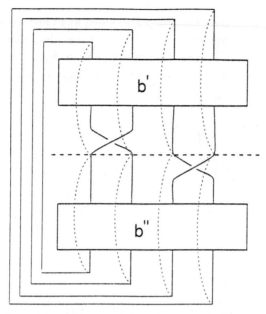

Fig. 2.8. Construction of brownian bridge for knots represented by B_3-braids.

the knot (link) K as a whole is characterized by the complexity $\eta = 0$ (i.e. the topological state of K "is close to trivial").

Introduce normalized generators of the group B_3

$$||\sigma_j^{\pm 1}|| = (\det \sigma_j^{\pm 1})^{-1} \sigma_j^{\pm 1}$$

to neglect the insignificant commutative factor dealing with norm of matrices σ_1 and σ_2. Now we can rewrite the power of Alexander invariant (Eq.(2.14)) in the form

$$\eta = [\#(+) - \#(-)] + \bar{\eta} \qquad (2.58)$$

where $\#(+)$ and $\#(-)$ are numbers of generators σ_{α_j} or $\sigma_{\alpha_j}^{-1}$ in a given braid and $\bar{\eta}$ is the power of the normalized matrix product $\prod_{j=1}^{N} ||\sigma_{\alpha_j}||$. The condition of brownian bridge implies $\eta = 0$ (i.e. $\#(+) - \#(-) = 0$ and $\bar{\eta} = 0$).

Theorem 7 *Take a set of knots obtained by closure of B_3-braids of length N with the uniform distribution over the generators. The conditional probability distribution $U(\bar{\eta}, m|N)$ for the normalized complexity $\bar{\eta}$ of*

Alexander polynomial invariant has the Gaussian behavior and is given by Eq.(2.54) *where* $k = \bar{\eta}$.

Proof. Write

$$||\sigma_1|| = T(t); \qquad ||\sigma_2|| = T^{-1}(t)S(t)T^{-1}(t) \qquad (2.59)$$

where $T(t)$ and $S(t)$ are the generators of the "t-deformed" group $PSL_t(2,\mathbb{Z})$

$$
\begin{aligned}
T(t) &= \begin{pmatrix} (-t)^{1/2} & 0 \\ 0 & (-t)^{-1/2} \end{pmatrix} \begin{pmatrix} 1 & (-t)^{-1} \\ 0 & 1 \end{pmatrix}; \\
T^{-1}(t) &= \begin{pmatrix} (-t)^{-1/2} & 0 \\ 0 & (-t)^{1/2} \end{pmatrix} \begin{pmatrix} 1 & -1 \\ 0 & 1 \end{pmatrix}; \qquad (2.60) \\
S(t) &= \begin{pmatrix} (-t)^{-1/2} & 0 \\ 0 & (-t)^{1/2} \end{pmatrix} \begin{pmatrix} 0 & 1 \\ -1 & 0 \end{pmatrix}
\end{aligned}
$$

The group $PSL_t(2,\mathbb{Z})$ preserves the relations of the group $PSL(2,\mathbb{Z})$ unchanged, i.e., $(T(t)S(t))^3 = S^4(t) = T(t)S^2(t)T^{-1}(t)S^{-2}(t) = 1$ (compare to Eq.(2.31)). Hence, if we construct the graph $C(\Gamma_t)$ for the group $PSL_t(2,\mathbb{Z})$ connecting the neighboring images of an arbitrary element from the fundamental domain, we ultimately come to the conclusion that the graphs $C(\Gamma_t)$ and $C(\Gamma)$ (fig.2.7) are topologically equivalent. This is the direct consequence of the fact that group B_3 is the central extension of $PSL(2,\mathbb{Z})$. It should be emphasized that the metric properties of the graphs $C(\Gamma_t)$ and $C(\Gamma)$ differ because of different embeddings of groups $PSL_t(2,\mathbb{Z})$ and $PSL(2,\mathbb{Z})$ into the complex plane.

Thus, the matrix product $\prod_{j=1}^{N} ||\sigma_{\alpha_j}||$ for the uniform distribution of braid generators is in one-to-one correspondence with the N-step random walk along the graph $C(\Gamma)$ (as it is explained in the proof of the Theorem 5). Its power coincides with the respective geodesics length along the backbone graph $C(\gamma)$. Taking into account Lemmas 2 and 3 we conclude that limit distribution of random walks on the group B_3 in terms of normalized generators (2.59) is given by Eq.(2.37) where k should be regarded as the power of the product $\prod_{\alpha=1}^{N} ||\sigma_{\alpha_j}||$. The statement of the Theorem follows now from the Collorary 1 $\square$.

2.3. Random Walks on Locally Free Groups

We aim at getting the asymptotics of conditional limit distributions of BB on the braid group B_n. For the case $n > 3$ it presents a problem which is unsolved yet. However we can estimate limit probability distributions of BB on B_n considering the limit distributions of random walks on the so-called "local groups" ([4,12,13]).

Definition 4 *The group* $\mathcal{LF}_{n+1}(d)$ *we call* **the locally free** *if the generators,* $\{f_1, \ldots, f_n\}$ *obey the following commutation relations:*

(a) *Each pair* (f_j, f_k) *generates the free subgroup of the group* $\mathcal{F}_n$ *if* $|j - k| < d$;

(b) $f_j f_k = f_k f_j$ *for* $|j - k| \geq d$

(Below we restrict ourselves to the case $d = 2$ where $\mathcal{LF}_{n+1}(2) \equiv \mathcal{LF}_{n+1}$).

Theorem 8 *The limit probability distribution for the* N-*step random walk* $(N \gg 1)$ *on the group* $\mathcal{F}_{n+1}$ *to have the minimal irreducible length* μ *is*

$$\mathcal{P}(\mu, N) \simeq \frac{\text{const}}{N^{3/2}} e^{-N/6} \mu \sinh \mu \exp\left(-\frac{3\mu^2}{2N}\right) \qquad (n = 3)$$

$$\mathcal{P}(\mu, N) \simeq \frac{1}{2\sqrt{14\pi N}} \exp\left\{-\frac{8}{7N}\left(\mu - \frac{3}{4}N\right)^2\right\} \qquad (n \gg 1)$$

$$(2.61)$$

Proof. We propose two independent approaches valid in two different cases: (1) for $n = 3$ and (2) for $n \gg 1$.

(1) The following geometrical image seems useful. Establish the one-to-one correspondence between the random walk in some n-dimensional Hilbert space $\mathcal{LH}^n(x_1, \ldots, x_n)$ and the random walk on the group $\mathcal{LF}_{n+1}$, written in terms of generators $\{f_1, \ldots, f_n^{-1}\}$. To be more specific, suppose that when a generator, say, f_j, (or f_j^{-1}) is added to the given word in $\mathcal{LF}_n$, the walker makes one unit step towards (backwards for f_j^{-1}) the axis $[0, x_j[$ in the space $\mathcal{LH}^n(x_1, \ldots, x_n)$.

Now the relations (a)-(b) of the Definition 4 could be reformulated in terms of metric properties of the space $\mathcal{LH}^n$. Actually, the relation (b) indicates that successive steps along the axes $[0, x_j[$ and $[0, x_k[$ $(|j - k| \geq 2)$ commute, hence the section (x_j, x_k) of the space $\mathcal{LH}^n$ is flat and has the Euclidean metric $dx_j^2 + dx_k^2$. Situation with the random trajectories in the sections $(x_j, x_{j\pm1})$ of the Hilbert space $\mathcal{LH}^n$ appears to be completely different. Here the steps of the walk obey the free group relations (a) and the walk itself is mapped to the walk on the Cayley tree. It is well known that Cayley tree can be uniformly embedded (without gaps and selfintersections) into the 3-pseudosphere which gives the representation of the non-Euclidean plane with the constant negative curvature. Thus, sections (x_j, x_{j+1}) have the metric of Lobachevskii plane which can be written in the form $\frac{1}{x_j^2}(dx_j^2 + dx_{j+1}^2)$.

For the group $\mathcal{LF}_4$ these arguments result in the following metric of appropriate space $\mathcal{LH}^{(3)}$

$$ds^2 = \frac{dx_1^2 + dx_2^2 + dx_3^2}{x_2^2} \qquad (2.62)$$

Actually, the space section (x_1, x_3) is flat whereas the space sections (x_1, x_2) and (x_2, x_3) have Lobachevskii plane metric. The noneuclidean (hyperbolic) distance between two points M' and M'' in the space $\mathcal{H}^3$ is defined as follows

$$\cosh \mu(M'M'') = 1 + \frac{1}{x_2(M')x_2(M'')} \sum_{i=1}^{3} (x_i(M') - x_i(M''))^2 \qquad (2.63)$$

where $\{x_1, x_2, x_3\}$ are the euclidean coordinates in the 3D-halfspace $x_2 > 0$ and μ is regarded as geodesics on a 4-pseudosphere (Lobachevskii space).

Some well known results concerning the limit behavior of random walks in spaces of constant negative curvature are reviewed in the next Section where solutions of the diffusion equations in the Lobachevskii plane and space are given by Eq.(2.84) and Eq.(2.86) correspondingly. Thus we can conclude that the distribution function for random walk in Lobachevskii space $\mathcal{P}_s(\mu, N)$ defined by Eqs.(2.86)-(2.89) gives also the probability for the N-letter random word (written in terms of uniformly distributed generators on $\mathcal{F}_4$) to have the primitive word of length μ (see Eq.(2.61)).

(2) For the group $\mathcal{LF}_{n+1}$ ($n \gg 1$) we extract the limit behavior of the distribution function exactly evaluating the volume of the maximal noncommutative subgroup of $\mathcal{LF}_{n+1}$.

Let $V_n(\mu)$ be the number of all nonequivalent primitive words of length μ on the group $\mathcal{LF}_{n+1}$. When $\mu \gg 1$, $V_n(\mu)$ has the following asymptotics:

$$V_n(\mu) = \text{const} \left[1 + 2\left(3 - \frac{4\pi^2}{n^2}\right)\right]^\mu; \qquad n \gg 1 \qquad (2.64)$$

To get Eq.(2.64) we write each primitive word W_p of length μ in the group $\mathcal{LF}_{n+1}$ in the so-called *normal order* (all f_{α_1} are different) similar to so-called "symbolic dynamics" used in consideration of chaotic systems

$$W_p = (f_{\alpha_1})^{m_1} (f_{\alpha_2})^{m_2} \cdots (f_{\alpha_s})^{m_s}, \tag{2.65}$$

where $\sum_{i=1}^{s} |m_i| = \mu \; (m_i \neq 0 \; \forall \; i; \; 1 \leq s \leq \mu)$ and sequence of generators f_{α_i} in Eq.(2.65) *for all* f_{α_i} satisfies the following local rules:

(i) If $f_{\alpha_i} = f_1$, then $f_{\alpha_{i+1}} \in \{f_2, f_3, \ldots f_{n-1}\}$;

(ii) If $f_{\alpha_i} = f_k \; (1 < k \leq n-1)$, then $f_{\alpha_{i+1}} \in \{f_{k-1}, f_{k+1}, \ldots f_{n-1}\}$;

(iii) If $f_{\alpha_i} = f_n$, then $f_{\alpha_{i+1}} = f_{n-1}$.

These local rules prescribe the enumeration of all distinct primitive words. If the sequence of generators in the primitive word W_p does not satisfy the rules (i)-(iii), we commute the generators in the word W_p up the normal order is restored. Hence, the normal order representation provides us with the unique coding of all nonequivalent primitive words in the group $\mathcal{LF}_{n+1}$.

Example 1. Take an arbitrary primitive word of length $\mu = 10$ in the group $\mathcal{LF}_8(2)$:

$$W_p = f_5^{-1} f_3 f_8 f_1^{-1} f_2 f_4 f_8 f_8 f_4 f_7 \tag{2.66}$$

To represent the word W_p in the "normal order" we have to push all generators with smaller indices to the left when it is allowed by the commutation relations of the locally free group $\mathcal{LF}_8(2)$. We get:

$$W_p = (f_1)^{-1} (f_3)^1 (f_2)^1 (f_5)^{-1} (f_4)^2 (f_8)^3 (f_7)^1 \tag{2.67}$$

(the "normal order" for this word is the sequence of used generators: $\{1, 3, 2, 5, 4, 8, 7\}$). To compute the number of different primitive words of length $\mu = 10$ with the same normal order as in Eq.(2.67), we have to sum up all the words like

$$W_p = (f_1)^{m_1} (f_3)^{m_2} (f_2)^{m_3} (f_5)^{m_4} (f_4)^{m_5} (f_8)^{m_6} (f_7)^{m_7} \tag{2.68}$$

under the condition $\sum_{i=1}^{7} |m_i| = 10; \; m_i \neq 0 \; \forall \; m_i \in [1, 7]$.

The calculation of the number of distinct primitive words, $V_n(\mu)$, of the given length μ is rather straightforward:

$$V_n(\mu) = \sum_{s=1}^{\mu} R(s) \sideset{}{'}\sum_{\{m_1,\ldots,m_s\}} \Delta \left[\sum_{i=1}^{s} |m_i| - \mu \right] \tag{2.69}$$

where $R(s)$ is the number of all distinct sequences of s generators taken from the set $\{f_1, \ldots, f_n\}$ and satisfying the local rules (i)-(iii) while the second sum gives the number of all possible representations of the primitive path of length μ *for the fixed sequence of generators* ("prime" means that the sum runs over all $m_i \neq 0$ for $1 \leq i \leq s$; Δ is the Kronecker Δ-function).

It should be mentioned that the local rules (i)-(iii) define the generalized Markov chain with the states given by the $n \times n$ coincidence matrix $\hat{T}_n$ where the rows and columns correspond to the generators $f_1, \ldots, f_n$:

$$\hat{T}_n(d) =$$

	f_1	f_2	f_3	f_4	$\cdots$	f_{n-1}	f_n
f_1	0	1	1	1	$\ldots$	1	1
f_2	1	0	1	1	$\ldots$	1	1
f_3	1	1	0	1	$\ldots$	1	1
f_4	0	1	1	0	$\ldots$	1	1
$\vdots$	$\vdots$	$\vdots$	$\vdots$	$\vdots$	$\ddots$	$\vdots$	$\vdots$
f_{n-1}	0	0	0	0	$\ldots$	0	1
f_n	0	0	0	0	$\ldots$	1	0

$$(2.70)$$

The matrix $\hat{T}_n(d)$ has rather simple structure: above the diagonal we put everywhere "1" and below diagonal we have $d - 1$ subdiagonals completely filled by "1"; in all other places we have "0" (in Eq.(2.70) it is shown the case with $d = 3$).

The number of all distinct normally ordered **sequences of words** of length s with allowed commutation relations is given by the following partition function

$$R_n(s, d) = \mathbf{v}_{\text{in}} \left[\hat{T}_n(d) \right]^s \mathbf{v}_{\text{out}} \qquad (2.71)$$

where

$$\mathbf{v}_{\text{in}} = (\overbrace{1 \, 1 \, 1 \, \ldots \, 1}^{n}) \quad \text{and} \quad \mathbf{v}_{\text{out}} = \left. \begin{pmatrix} 1 \\ 1 \\ 1 \\ \vdots \\ 1 \end{pmatrix} \right\} n \qquad (2.72)$$

Supposing that the main contribution in Eq.(2.69) results from $s \gg 1$ we take for $R_n(s)$ the following asymptotic expression

$$R_n(s)\Big|_{s \gg 1} = (\lambda_n^{max})^s \qquad (2.73)$$

where λ_n^{max} is the highest eigenvalue of the matrix $\hat{T}_n$ $(n \gg 1)$.

Simple though tedious calculations give the following value for the highest eigenvalue λ_n^{max} for $n \gg 1$

$$\lambda_n^{max} = 3 - \frac{4\pi^2}{n^2} + O\left(\frac{1}{n^3}\right) \qquad (2.74)$$

The remaining sum in Eq.(2.69) is independent of $R(s)$, so its calculation is trivial:

$$\sum_{\{m_1,\ldots,m_s\}}' \Delta \left[\sum_{i=1}^{s} |m_i| - \mu\right] = 2^s \frac{(\mu - 1)!}{(s-1)!(\mu - s)!} \qquad (2.75)$$

Substituting Eq.(2.69) for Eqs.(2.74) and (2.75) and evaluating the sum over s we arrive at Eq.(2.64). The value $V_n(\mu, d)$ is growing exponentially fast with μ and the "speed" of this grows is clearly represented by the fraction

$$z_{\text{eff}} - 1 = \frac{V_n(\mu + 1)}{V_n(\mu)}\Big|_{\mu \gg 1} \simeq 7 - \frac{8\pi^2}{n^2} \qquad (2.76)$$

where z_{eff} is the coordinational number of effective tree associated with the locally free group.

Thus, the random walk on the group $\mathcal{LF}_{n+1}$ can be viewed as follows. Take the *free* group Γ_n with generators $\{\tilde{f}_1, \ldots, \tilde{f}_n\}$ where all $\tilde{f}_i$ $(1 \le i \le n)$ do not commute. The group Γ_n has a structure of $2n$-branching Cayley tree, $C(\Gamma_n)$, where the number of distinct words of length μ is equal to $\tilde{V}_n(\mu)$,

$$\tilde{V}_n(\mu) = 2n(2n-1)^{\mu-1} \qquad (2.77)$$

The graph $C(\mathcal{LF}_{n+1})$ corresponding to the group $\mathcal{LF}_{n+1}$ can be constructed from the graph $C(\Gamma_n)$ in accordance with the following recursion procedure:

a) Take the root vertex of the graph $C(\Gamma_n)$ and consider all vertices on the distance $\mu = 2$. Identify those vertices which correspond to the equivalent words in group $\mathcal{LF}_{n+1}$. (See example in fig.2.9).

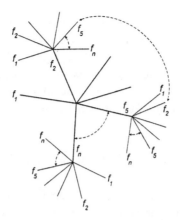

Fig. 2.9. The vertices A and B should be glued because they represent one and the same word in group $\mathcal{LF}_{n+1}$.

b) Repeat this procedure taking all vertices at the distance $\mu = (1, 2, \ldots)$ and "gluing" them at the distance $\mu + 2$ according to Definition 4.

By means of the described procedure we raise a graph which in average has $z_{\text{eff}} - 1$ distinct branches leading to the "next coordinational sphere". Thus this graph coincides (in average) with z_{eff}-branching Cayley tree.

Although the local structure of the graph $C(\mathcal{LF}_{n+1})$ is very complex, Eq.(2.76) enables us to find the asymptotics of the random walk on the graph $C(\mathcal{LF}_{n+1})$. Once having z_{eff}, we can write down the master equation for the probability $\mathcal{P}(\mu, N)$ to find the walker at the distance μ from the origin after N random steps on the graph $C(\mathcal{LF}_{n+1})$

$$\mathcal{P}(\mu, N+1) = (1 - \pi)\, P(\mu - 1, N) + \pi\, P(\mu + 1, N) \qquad (\mu \geq 2) \quad (2.78)$$

where

$$\pi = \frac{1}{z_{\text{eff}}}$$

The recursion relation (2.78) coincides with the equation describing the random walk on the halfline with the drift from the origin. Taking into account this analogy we can complete the Eq.(2.78) by the boundary conditions

$$\mathcal{P}(\mu = 1, N+1) = P(\mu = 0, N) + \pi\, P(\mu = 2, N)$$
$$\mathcal{P}(\mu = 0, N+1) = \pi P(\mu = 1, N) \qquad\qquad (2.79)$$
$$\mathcal{P}(\mu, N = 0) = \delta_{\mu,0}$$

It is noteworthy that these equations are written just for the Cayley tree with z_{eff} branches. Let us repeat once more that the actual structure of the graph corresponding to the group $\mathcal{LF}_n(d)$ is much more complex, thus Eqs.(2.78)–(walk2) should be regarded as an approximation. However the exact form of boundary conditions does not influence the asymptotic solution of Eq.(2.78) in vicinity of the maximum of the distribution function:

$$P(\mu, N) \simeq \frac{1}{2\sqrt{2\pi(z_{\text{eff}} - 1)N}} \exp\left\{ -\frac{z_{\text{eff}}^2}{8(z_{\text{eff}} - 1)N} \left(\mu - \frac{z_{\text{eff}} - 2}{z_{\text{eff}}}N \right)^2 \right\}$$

Thus we obtain the desired distribution function (Eq.(2.61)) for the primitive word length for the random walk on the group $\mathcal{LF}_{n+1}$ $\square$.

Corollary 2 *The Eq.(2.61) gives the estimation from below for the limit distribution of the primitive words on the group B_n for $n \gg 1$.*

We find further investigation of the random walks on the groups $\mathcal{LF}_{n+1}(d)$ for different values of d very perspective. It should give insight for consideration of random walk statistics on "partially commutative groups" moreover it could be regarded as a natural model for the problem of limit distributions on the group of coloured braids.

2.4. Brownian Bridges on Lobachevskii Plane and Products of Noncommutative Random Matrices

Let us put aside for some time topological problems and consider the random walks on the plane with the constant negative curvature (the Lobachevskii plane). The present Section is mainly illustrative. We begin by showing that the brownian bridge condition in the Lobachevskii plane (space) nullifies the effect of the curvature turning the corresponding limit distribution function to the Gaussian one with zero mean. Then we reformulate the problem of the N-step random walk in Lobachevskii plane in terms of the N-step Markov chain with the states in ensemble of equally distributed independent noncommutative random matrices and ask for the limit behavior of corresponding brownian bridge. In other words, we look for the limit distribution of the Lyapunov exponent of first m matrices in the product of N random matrices *under the condition that the whole product is identical to unit matrix.*

The rigorous analysis of the conditional limit distributions for the Lyapunov exponents of the products of the 2×2 unimodular random matrices is presented in Appendix A, where the paper [9] is reproduced with some minor reductions.

Recall that the distribution function, $P(\mathbf{r}, t)$, for the free random walk in D-dimensional Euclidean space obeys the standard heat equation:

$$\frac{\partial}{\partial t} P(\mathbf{r}, t) = \mathcal{D}\Delta P(\mathbf{r}, t)$$

with the diffusion coefficient $\mathcal{D} = \frac{1}{2D}$ and appropriate initial and normalization conditions

$$P(\mathbf{r}, t = 0) = \delta(\mathbf{r})$$

$$\int P(\mathbf{r}, t) d\mathbf{r} = 1$$

Correspondingly, the diffusion equation for the scalar density $P(\mathbf{q}, t)$ of the free random walk on a Riemann manifold reads (see [26] for instance)

$$\frac{\partial}{\partial t} P(\mathbf{q}, t) = \mathcal{D} \frac{1}{\sqrt{g}} \frac{\partial}{\partial q_i} \left(\sqrt{g} \left(g^{-1} \right)_{ik} \frac{\partial}{\partial q_k} \right) P(\mathbf{q}, t) \qquad (2.80)$$

where

$$P(\mathbf{q}, t = 0) = \delta(\mathbf{q})$$

$$\int \sqrt{g} P(\mathbf{q}, t) d\mathbf{q} = 1 \qquad (2.81)$$

and g_{ik} is the metric tensor of the manifold; $g = \det g_{ik}$.

Eq.(2.80) has been subjected to thorough analysis for the manifolds of the constant negative curvature. Below we reproduce the corresponding solutions for the best known cases: for 2D- and 3D–Lobachevskii spaces (often referred to as 3– and 4–pseudospheres) labelling them by indices "p" and "s" for 2D– and 3D–cases correspondingly.

For the Lobachevskii plane one has

$$\|g_{ik}\| = \left\| \begin{matrix} 1 & 0 \\ 0 & \sinh^2 \mu \end{matrix} \right\| \qquad (2.82)$$

where μ stands for the geodesics length on 3-pseudosphere. The corresponding diffusion equation now reads

$$\frac{\partial}{\partial t} P_p(\mu, \varphi, t) = \mathcal{D} \left(\frac{\partial^2}{\partial \mu^2} + \coth \mu \frac{\partial}{\partial \mu} + \frac{1}{\sinh^2 \mu} \frac{\partial^2}{\partial \varphi^2} \right) P_p(\mu, \varphi, t) \qquad (2.83)$$

The solution of Eq.(2.83) is believed to have the following form

$$
P_p(\mu, t) = \frac{e^{-\frac{tD}{4}}}{4\pi\sqrt{2\pi(tD)^3}} \int_\mu^\infty \frac{\xi \exp\left(-\frac{\xi^2}{4tD}\right)}{\sqrt{\cosh\xi - \cosh\mu}} d\xi
$$

$$
\simeq \frac{e^{-\frac{tD}{4}}}{4\pi tD} \left(\frac{\mu}{\sinh\mu}\right)^{1/2} \exp\left(-\frac{\mu^2}{4tD}\right)
$$

(2.84)

For the Lobachevskii space the corresponding metric tensor is

$$
||g_{ik}|| = \left\| \begin{matrix} 1 & 0 & 0 \\ 0 & \sinh^2\mu & 0 \\ 0 & 0 & \sinh^2\mu\sin^2\theta \end{matrix} \right\|
$$

(2.85)

Substituting Eq.(2.80) for Eq.(2.85) we have

$$
P_s(\mu, t) = \frac{e^{-tD}}{8\pi\sqrt{\pi(tD)^3}} \frac{\mu}{\sinh\mu} \exp\left(-\frac{\mu^2}{4tD}\right)
$$

(2.86)

For the first time this spherically symmetric solution of the heat equation (Eq.(2.80)) in the Lobachevskii space was received in [28,27].

In our opinion one fact must be given our attention. The distribution functions $P_p(\mu, t)$ and $P_s(\mu, t)$ give the probabilities to find the random walk starting at the point $\mu = 0$ after time t in some *specific* point located at the distance μ in corresponding noneuclidean space. The probability to find the terminal point of a random walk after time t *somewhere* at the distance μ is

$$
\mathcal{P}_{p,s}(\mu, t) = P_{p,s}(\mu, t)\mathcal{N}_{p,s}(\mu)
$$

(2.87)

where

$$
\mathcal{N}_p(\mu) = \sinh\mu
$$

(2.88)

is the perimeter of circle of radius μ on the Lobachevskii plane and

$$
\mathcal{N}_s(\mu) = \sinh^2\mu
$$

(2.89)

is the area of sphere of radius μ in the Lobachevskii space.

The difference between $P_{p,s}$ and $\mathcal{P}_{p,s}$ is insignificant in euclidean geometry, whereas in the noneuclidean space it becomes dramatic because of the consequences of the behavior of brownian bridges in spaces on constant negative curvature.

Using the definition of the brownian bridge, let us calculate the probabilities to find the N-step random walk (starting at $\mu = 0$) after first t steps at the distance μ in the Lobachevskii plane (space) under the condition that it returns to the origin on the last step. These probabilities are $(N \to \infty)$

$$
\begin{aligned}
\mathcal{P}_p(\mu, t|0, N) &= \frac{P_p(\mu, t)\mathcal{P}_p(\mu, N - t)}{P_p(0, t)} \\
&\simeq \frac{N}{4\pi \mathcal{D} t(N - t)} \mu \exp\left\{-\frac{\mu^2}{4\mathcal{D}}\left(\frac{1}{t} + \frac{1}{N - t}\right)\right\} \\
\mathcal{P}_s(\mu, t|0, N) &= \frac{P_s(\mu, t)\mathcal{P}_s(\mu, N - t)}{P_s(0, t)} \\
&\simeq \frac{N^{3/2}}{8\pi t^{3/2}(N - t)^{3/2}} \mu^2 \exp\left\{-\frac{\mu^2}{4\mathcal{D}}\left(\frac{1}{t} + \frac{1}{N - t}\right)\right\}
\end{aligned}
\tag{2.90}
$$

Hence we come to the standard Gaussian distribution function with zero mean.

Equations (2.90) describing the random walk on the Riemann surface of constant negative curvature have direct application to the conditional distributions of Lyapunov exponents for products of some noncommutative matrices. Let us consider the first of Eqs.(2.90). Changing the variables

$$
\mu = \ln \frac{1 + |z|}{1 - |z|}; \quad \varphi = \arg z
$$

where $z = x + iy$; $\bar{z} = x - iy$ we map the 3–pseudosphere (μ, φ) onto the unit disk $|z| < 1$ known as the Poincare representation of the Lobachevskii plane. The corresponding conformal metric reads

$$
dl^2 = \frac{4\, dz d\bar{z}}{\left(1 - |z|^2\right)^2}
$$

Using the conformal transform $z = \dfrac{1 + iw}{1 - iw}$ we recover the so-called Klein representation of Lobachevskii plane, where

$$
dl^2 = -\frac{4\, dw d\bar{w}}{(w - \bar{w})^2}
$$

and the model is defined in $\text{Im} w > 0$ $(w = u + iv; \bar{w} = u - iv)$.

The following relations can be verified using conformal representations of the Lobachevskii plane metric (see, for instance, [29]). The fractional group of motions of Lobachevskii plane is isomorphic to:

(i) the group $SU(1,1)/\pm 1 \equiv PSU(1,1)$ in the Poincare model;

(ii) the group $SL(2,\mathbf{R})/\pm 1 \equiv PSL(2,\mathbf{R})$ in the Klein model.

Moreover, it is known (see, for example, [30]) that the Lobachevskii plane H can be identified with the group $SL(2,\mathbf{R})/SO(2)$. This relation enables us to resolve (at least qualitatively) the following problem. Take the brownian bridge on the group $\mathcal{H} = SL(2,\mathbf{R})/SO(2)$, i.e. demand the products of N independent random matrices $\widehat{\mathcal{M}}_k \in \mathcal{H}$ ($0 \leq k \leq N$) to be identical to the unit matrix. Consider the limit distribution of the Lyapunov exponent, $\hat{\delta}$, for the first m matrices in that products. To have a direct mapping of this problem on the random walk in the Lobachevskii plane, write the corresponding stochastic recursion equation for some vector $\mathbf{W}_k = \begin{pmatrix} u_k \\ v_k \end{pmatrix}$

$$\mathbf{W}_{k+1} = \widehat{\mathcal{M}}_k \mathbf{W}_k; \qquad \mathbf{W}_0 = \begin{pmatrix} 1 \\ 1 \end{pmatrix} \tag{2.91}$$

where $\mathcal{M}_k \in \mathcal{H}$ for all $k \in [0, N]$. The BB–condition means that

$$\mathbf{W}_N = \mathbf{W}_0 \quad \text{for } N \gg 1 \tag{2.92}$$

Let us consider the simplest case

$$\widehat{\mathcal{M}}_k = 1 + \widehat{M}_k; \qquad \text{norm}[\widehat{M}_k] \ll 1 \tag{2.93}$$

In this case the discrete dynamic equation (2.91) can be replaced by the differential one. Its stationary measure is determined by the corresponding Fokker-Plank equation (2.80). The Lyapunov exponent, $\hat{\delta}$ of product of random matrices $\widehat{\mathcal{M}}$ coincides with the length of geodesics in the Klein representation of the Lobachevskii plane. Hence, under the conditions (2.92), (2.93) we have for $\hat{\delta}$ the usual Gaussian distribution coinciding with the first of Eq.(2.90). Without the BB–condition (i.e. for "open walks") we reproduce the standard Fürstenberg behavior [8].

Although this consideration seems rather crude (for details see Appendix A), it clearly shows the origin of the main result:

The "brownian bridge" condition for random walks in space of constant negative curvature makes the space "effectively flat" turning the corresponding limit probability distribution for random walks to the ordinary central limit distribution.

The question whether this result is valid for the case of the random walk in noneuclidean spaces of nonconstant negative curvature still remains.

2.5. Remarks and Conclusions

Let us mention the simple geometrical meaning of results concerning the random walks on B_3:

(a) The limit distribution of the shortest noncontractible word for the random walk on the group is rather rough characteristic being not very sensitive to the local group relations.

(b) Comparing the random walks on the free (Γ_2) and braid (B_3) groups, we can see that the presence of the Yang-Baxter-type relations change effectively only the number of corresponding branches (i.e. the effective curvature of corresponding non-Euclidean space, in which graph $\tilde{\Gamma}$ could be embedded). In particular, the "effective coordinational number", z, of the backbone graph, $C(\gamma)$, corresponding to the group $PSL(2,\mathbb{Z})$ is $z = 3$ while for the graph, representing the free group, $z = 4$.

(c) It should be noted that the "brownian bridge" condition of the random walk on the local free groups (as well as on the free one) fully reimburses for the "drift from the origin" turning the corresponding limit probability distribution into the Gaussian one **with zero mean** if the distribution of group generators is uniform. We believe that this property is common for the random walks on the noncommutative groups. Anyway, the said behavior has been recently described in several works [2,9,24] (see also Chapter 4).

Finally we would like to introduce some conjectures which naturally generalize our consideration.

Conjecture 1 *The complexity η of any known algebraic invariants* (Alexander, Jones, HOMFLY) *for the knot represented by the B_n-braid of length N with the uniform distribution over generators has the following limit behavior:*

$$P(\eta, N) \sim \frac{\text{const}}{N^{3/2}} \eta \exp\left(-\alpha(n)N + \beta(n)\eta - \frac{\eta^2}{\delta(n)N}\right) \qquad (2.94)$$

where $\alpha(n)$, $\beta(n)$, $\delta(n)$ are numerical constants depending on n only.

Conjecture 2 *The knot complexity η in ensemble of brownian bridges from the group B_n shown in fig.2.8 has Gaussian distribution, where*

$$\langle \overline{\eta} \rangle = 0; \qquad \langle \overline{\eta}^2 \rangle = \frac{1}{2}\delta(n)N \tag{2.95}$$

These conjectures are to be proven yet. The main idea is to employ the relation between the knot complexity η, the length of the shortest non-contractible word and the length of geodesics on some hyperbolic manifold.

Conjecture 3 *Take the product of N uniformly distributed independent matrices, $\prod_{\alpha=1}^{N} \hat{f}_{\alpha_j}$, from the set $\{\hat{f}_1, \ldots, \hat{f}_n^{-1}\}$, ($\hat{f}_j$ gives the matrix representation of generators f_j of the group $\mathcal{F}_{n+1}$ for all j). Then the value $\mu = \ln \left| \text{Trace} \prod_{j=1}^{N} \hat{f}_{\alpha_j} \right|$ in the limit $1 \ll n \ll N$ has the Gaussian probability distribution with non-zero mean.*

The matrix representation of the group $\mathcal{LF}_{n+1}$ resembles the Magnus representation of the braid group generators Eq.(2.13) (but without the Yang-Baxter relations) where the block A is as follows[††]:

$$A = \begin{pmatrix} 1 & 0 & 0 \\ 2 & 1 & 2 \\ 0 & 0 & 1 \end{pmatrix} \tag{2.96}$$

Define the Lyapunov exponent, $\lambda(N)$, of the product $\prod_{j=1}^{N} \hat{f}_{\alpha_j}$ as follows: $\lambda(N) \overset{def}{=} \ln |\Lambda(N)|$, where $\Lambda(N)$ is the highest eigenvalue of the product in question. In the limit $N \to \infty$ ($n = $ const) we can rewrite $\lambda(N)$ as

$$\lambda(N) \simeq \ln \left| \text{Trace} \prod_{j=1}^{N} \hat{f}_{j_\alpha} \right| \tag{2.97}$$

On the other hand, $\lambda(N)$ is proportional to the length of the shortest non-contractible word (the length of "geodesics"), μ, for the random walk on the group $\mathcal{F}_n$. Using Eq.(2.61) we obtain the posed conjecture. However schematic such consideration may seem, it can be supported by the following arguments. The relation

$$|Tr(N)| = 2\cosh\frac{l_p}{2} \tag{2.98}$$

[††]The representation Eq.(2.13) satisfies the Hecke algebra relations Eq.(2.8) with $t = -1$.

establishes the connection between the trace and the length of the periodic orbit, l_p, on some group. It is known (see, for instance, [25]) that in average the number of periodic orbits N_p of the length l_p is proportional to $l_p^{-1} \exp(l_p/2)$ for the hyperbolic groups. At the same time the volume $V_n(\mu)$ of the group $\mathcal{LF}_{n+1}$ is growing as const $\times \, 7^\mu$. Comparing N_p and $V_n(\mu)$ for $\mu \gg 1$ and $l_p \gg 1$ we get

$$\left. \frac{\mu}{l_p} \right|_{l_p \to \infty} = \frac{1}{2\ln 7} + O\left(\frac{\ln l_p}{l_p} \right)$$

Supposing that l_p is proportional to the length of geodesics, μ, and taking into account the distribution function Eq.(2.61), we return to the third conjecture.

References

1. S.K.Nechaev, A.Yu. Grosberg, A.M.Vershik, J. Phys. (A): Math. Gen. ?? (1996), ??

2. S. Nechaev, Ya.G. Sinai, Bol. Soc. Bras. Mat., 21 (1991), 121

3. H. Kesten, Trans. Amer. Math. Soc., 92 (1959), 336

4. A.M. Vershik, in *Topics in Algebra*, 26, pt.2 (1990), 467, (Banach Center Publication, PWN Publ., Warszawa); Proc. Am. Math. Soc., 148 (1991), 1

5. P. Chassaing, G. Letac, M. Mora, in *Probanility Measures on Groups*, Lect. Not. Math., 1064 (1983)

6. Ya.G. Sinai, *Introduction to Ergodic Theory*, (Princeton Univ. Press: Princeton, NJ, 1977)

7. M. Gutzwiller, *Chaos in Classical and Quantum Mechanics* (Springer: N.Y., 1990)

8. H. Fürstenberg, Trans. Am. Math. Soc, 198 (1963), 377; V. Tutubalin, Prob. Theor. Appl. (in Russian), 10 (1965), 15

9. L. Koralov, S. Nechaev, Ya. Sinai, Prob. Theor. Appl. 38 (1993), 331 (in Russian)

10. E. Helfand, D.S. Pearson, J. Chem. Phys., 79 (1983), 2054; M. Rubinstein, E. Helfand, J. Chem. Phys., 82 (1985), 2477

11. A. Khokhlov, S. Nechaev, Phys. Lett. A112 (1985), 156

12. S. Nechaev, A. Vershik, J. Phys. (A): Math. Gen., 27 (1994), 2289

13. J. Desbois, S. Nechaev *Statistical Mechanics of Braided Markov Chains: I. Analytic Methods and Numerical Simulations*, Preprint IPNO/TH 96-20 (Submitted to J. Stat. Phys.)

14. S. Nechaev, A. Semenov, M. Koleva, Physica, A140 (1987), 506

15. A. Mehta, J. Needs, Thouless, Europhys. Lett., ?? (1988), ??

16. V.F.R. Jones, Bull. Am. Math. Soc., 12 (1985), 103

17. A.M. Vershik, S. Kerov, Dokl. Ac. Nauk SSSR, 301 (1988), 777 (In Russian)
18. L.H. Kauffman, H. Saleur, Comm. Math. Phys., 141 (1991), 293
19. Y. Akutsu, T.K. Deguchi, Phys. Rev. Lett., 67 (1991), 777
20. Y. Akutsu, M. Wadati, J. Phys. Soc. Jap., 56 (1987),
21. J. Birman, *Knots, Links and Mapping Class Groups*, Ann. Math. Stud., 82, (Princeton Univ. Press: Princeton, NJ, 1976)
22. A. Grosberg, S. Nechaev, Europhys. Lett., 20 (1992), 613
23. Mumford, *Tata Lectures on Theta, I, II*, Progress in Mathematics, 28, 34, (Boston, MA: Birkhauser, 1983)
24. A. Letchikov, Prob. Theor. Appl. (in Russian) to be published
25. E. Bogomolny, F. Leyvarz, C. Schmit, preprint IPNO/TH 94-43, to appear in Comm. Math. Phys.
26. J.Zinn-Justin, *Quantum Field Theory and Critical Phenomena* (Clarendon: Oxford, 1989)
27. M.E. Gerzenshtein, V.B. Vasiljev, Prob. Theor. Appl., 4 (1959), 424 (in Russian)
28. F.I. Karpelevich, V.N. Tutubalin, M.G. Shour, Prob. Theor. Appll., 4 (1959), 432 (in Russian)
29. B.A. Dubrovin, S.P. Novikov, A.T. Fomenko *Modern Geometry*, (Nauka: Moscow, 1979)
30. A. Terras *Harmonic Analysis on Symmetric Spaces and Applications I*, (Springer: Heidelberg)

CHAPTER 3

CONFORMAL METHODS IN STATISTICS
OF ENTANGLED RANDOM WALKS

3.1. Introduction: Random Walk with Topological Constraints

The last few years have been marked by considerable progress in understanding the relationship between Chern-Simons topological field theory, construction of algebraic knot and link invariants and conformal field theory (see, for review, [1,2]).

Although the general concepts have been well elaborated in the field-theoretic context, their application in the related areas of mathematics and physics, such as, for instance, probability theory and statistical physics of chain-like objects is highly limited. This state of affairs can be accounted for two facts:

(a) There is a problem of communication, i.e. the languages used by specialists in topological field theory and probability theory are completely different on the first sight;

(b) Physical systems give no evidence of reflection of these ideas in simple geometrical examples.

The present Chapter is mainly concerned with the conformal methods in statistical analysis which allow us to correlate problems discussed in Chapters 1 and 2 and the limit distributions of random walks on multiconnected Riemann surfaces.

To be more specific, we show on the level of differential equations how simple geometrical methods can be applied to construction of noncommutative topological invariants. The latter might serve as nonabelian generalizations of the Gauss linking numbers for the random walks on multi-punctured Riemann surfaces. We also study the connection between the topological properties of random walks on the double punctured plane and behavior of four-point correlation functions in the conformal theory with central charge $C = -2$. The developed approach is applied to the investigation of statistics of 2D–random walks with multiple topological constraints. For instance, the methods presented here allow us to extract nontrivial critical exponents for the contractible (i.e., unentangled) random walks in the regular lattices of obstacles. Some of our findings support conjectures of Chapters 1 and 2 and have direct application in statistics of strongly entangled polymer chains (see Chapter 4).

3.2. Construction of Nonabelian Connections for Γ_2 and $PSL(2,\mathbb{Z})$ from Conformal Methods

We analyse the random walk of length L with the effective elementary step a ($a \equiv 1$) on the complex plane $z = x + iy$ with two points removed. Suppose the coordinates of these points being M_1 ($z_1 = (0,0)$) and M_2 ($z_2 = (c,0)$) ($c \equiv 1$). Such choice does not indicate the loss of generality because by means of simultaneous rescaling of the effective step, a, of the random walk and of the distance, c, between the removed points we can always obtain of any arbitrary values of a and c.

Consider the closed paths on z and attribute the generators g_1, g_2 of some group G to the turns around the points M_1 and M_2 if we move along the path in the clockwise direction (we apply g_1^{-1}, g_2^{-1} for counter-clockwise move)—see fig.3.1.

The question is: what is the probability $P(\mu, L)$ for the random walk of length L on the plane z to form a closed loop with the shortest noncontractible word written in terms of generators $\{g_1, g_2, g_1^{-1}, g_2^{-1}\}$ to have the length μ (see also Chapter 2).

Let the distribution function $P(\mu, L)$ be formally written as a path integral with a Wiener measure

$$P(\mu, L) = \frac{1}{\mathcal{Z}} \int \dots \int \mathcal{D}\{z\} \, \exp\left\{-\frac{1}{a^2} \int_0^L \left(\frac{dz(s)}{ds}\right)^2 ds\right\}$$

$$\times \delta\left[W\{g_1, g_2, g_1^{-1}, g_2^{-1}|z\} - \mu\right]$$

(3.1)

where $\mathcal{Z} = \int P(\mu, L)d\mu$ and $W\{\dots|z\}$ is the length of the shortest word

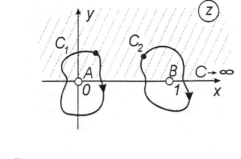

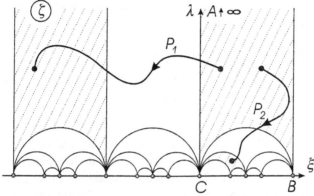

Fig. 3.1. (a)—the double punctured complex plane z with two basis loops C_1 and C_2 enclosing points M_1 and M_2; (b)—the universal covering ζ with fundamental domain corresponding to free group Γ_2. The contours P_1 and P_2 are the images of the loops C_1 and C_2.

on G as a functional of the path on the complex plane.

Conformal methods enable us construct the connection and the topological invariant W for the given group as well as to rewrite Eq.(3.1) in a closed analytic form which is solvable at least in the limit $L \to \infty$.

Let $\zeta(z)$ be the conformal mapping of the double punctured plane $z = x + iy$ on the universal covering $\zeta = \xi + i\lambda$. The Riemann surface ζ is constructed in the following way. Make three cuts on the complex plane z between the points M_1 and M_2, between M_2 and (∞) and between (∞) and M_1 along the line $\mathrm{Im}z = 0$. These cuts separate the upper ($\mathrm{Im}z > 0$) and lower ($\mathrm{Im}z < 0$) halfplanes of z. Now perform the conformal transform of the halfplane $\mathrm{Im}z > 0$ to the fundamental domain of the group $G\{g_1, g_2\}$— the curvilinear triangle lying in the halfplane $\mathrm{Im}\zeta > 0$ of the plane ζ. Each fundamental domain represents the Riemann sheet corresponding to the fibre bundle above z. The whole covering space ζ is the unification of all such Riemann sheets.

We distinguish between different topological states of the path with respect to the removed points on z by the topological invariants constructed from the conformal mapping described above. Shown in fig.3.1a are two closed contours C_1 and C_2 starting and ending in some arbitrary point z_0 on z. These contours belong to the same homology class but have different classes of homotopy.

Statement 1 *The coordinates of initial and final points of any trajectory on universal covering ζ determine* ([5]):

(a) *The coordinates of corresponding points on z;*

(b) *The homotopy class of any path on z. In particular, the contours on ζ are closed if and only if $W\{g_1, g_2 | z\} \equiv 1$, i.e. they belong to the trivial homotopy class.*

Coordinates of ends of the trajectory on universal covering ζ can be used as the topological invariant for the path on double punctured plane z with respect to the action of the group G. It should be emphasized that this invariant reflects the non-abelian character of entanglements correctly and is *complete* for our problem.

Thus, we characterize the topological invariant, $\mathrm{Inv}(C)$, of some closed directed path C starting and ending in an arbitrary point $z_0 \neq \{z_1, z_2, \infty\}$ on the plane z by the coordinates of the initial, $\zeta_{in}(z_0)$, and final, $\zeta_{fin}(z_0)$, points of the corresponding contour P in the covering space ζ. The contour P connects the images of the point z_0 on the different Riemann sheets. Write $\mathrm{Inv}_{(z)}(C)$ as a full derivative along the contour C:

$$\mathrm{Inv}_{(z)}(C) \stackrel{def}{=} \zeta_{in} - \zeta_{fin} = \oint_C \frac{d\zeta(z)}{dz} dz \qquad (3.2)$$

The physical interpretation of the derivative $\dfrac{d\zeta(z)}{dz}$ is very straightforward. Actually, the invariant, $\mathrm{Inv}(C)$, can be associated with the flux through the contour C on the plane (x, y):

$$\mathrm{Inv}(C) \equiv \mathrm{Inv}_{(x,y)}(C) = \oint_C \nabla\zeta(x,y)\mathbf{n}d\mathbf{r} = \oint_C \boldsymbol{\nu} \times \nabla\zeta(x,y)\mathbf{v}(s)ds \qquad (3.3)$$

where: $\mathbf{n}$ is the unit vector normal to the curve C, $d\mathbf{r} = \mathbf{e}_x dx + \mathbf{e}_y dy$ on the plane (x, y); $\mathbf{v}(s) = \dfrac{d\mathbf{r}}{ds}$ denotes the "velocity" along the trajectory; and ds stands for the differential path length. Simple transformations used in Eq.(3.3) are:

(a) $\mathbf{n}d\mathbf{r} = \mathbf{e}_x dy - \mathbf{e}_y dx = d\mathbf{r} \times \nu$;

(b) $\nabla\zeta(x, y)(d\mathbf{r} \times \nu) = (\nu \times \nabla\zeta(x, y))d\mathbf{r}$, where $\nu = (0, 0, 1)$ is the unit vector normal to the plane (x, y).

The vector product

$$\mathbf{A}(x, y) = \nu \times \nabla\zeta(x, y) \tag{3.4}$$

can be considered a non-abelian generalization of the vector potential of a solenoidal "magnetic field" normal to the plane (x, y) and crossing it in the points (x_1, y_1) and (x_2, y_2). Thus, $\mathbf{A}$ defines the *flat connection* of the double punctured plane z with respect to the action of the group G.

It is easy to show how the basic formulae (3.2) and (3.3) transform in case of commutative group $G_{comm}\{g_1, g_2\}$ which distinguishes only the classes of homology of the contour C with respect to the removed points on the plane. The corresponding conformal transform is performed by the function $\zeta(z) = \ln(z - z_1) + \ln(z - z_2)$. This immediately gives the abelian connection and the Gauss linking number as a topological invariant:

$$\mathbf{A}(\mathbf{r}) = \nu \times \sum_{j=\{1,2\}} \frac{\mathbf{r} - \mathbf{r}_j}{|\mathbf{r} - \mathbf{r}_j|^2};$$

$$\mathrm{Inv}(C) = \oint_C \mathbf{A}(\mathbf{r})d\mathbf{r} = \sum_{j=\{1,2\}} \oint_C \frac{(y - y_j)dx - (x - x_j)dy}{(x - x_j)^2 + (y - y_j)^2}$$

$$= 2\pi(n_1 + n_2)$$

where n_1 and n_2 are the winding numbers of the path C around the points M_1 and M_2 of the plane (x, y).

Substituting Eq.(3.1) written in the Euclidean coordinates (x, y) for Eq.(3.3) and using the Fourier transform for the δ-function, we can rewrite equation (3.1) as follows

$$P(\mu, L) = \frac{1}{2\pi} \int_{-\infty}^{\infty} e^{-iq\mu} P(q, L)dq \tag{3.5}$$

where

$$P(q, L) = \frac{1}{Z} \int \cdots \int \mathcal{D}\{\mathbf{r}\} \exp \left\{ -\frac{1}{a^2} \int\limits_0^L \left(\left(\frac{d\mathbf{r}(s)}{ds}\right)^2 - iq\mathbf{A}(\mathbf{r})\frac{d\mathbf{r}(s)}{ds} \right) ds \right\}$$

(3.6)

The function $P(q, L)$ coincides with the Green function $P(\mathbf{r}_0, \mathbf{r} = \mathbf{r}_0, q, L)$ of the non-stationary Schrödinger-like equation for the free particle motion in a "magnetic field" with the vector potential (3.4):

$$\frac{\partial}{\partial L} P(\mathbf{r}_0, \mathbf{r}, q, L) - \left(\frac{1}{2a}\nabla - iq\mathbf{A}(\mathbf{r}) \right)^2 P(\mathbf{r}_0, \mathbf{r}, q, L) = \delta(L)\delta(\mathbf{r} - \mathbf{r}_0) \quad (3.7)$$

where q plays a role of a "charge" and the magnetic field is considered transversal, i.e. $\mathrm{rot}\,\mathbf{A}(\mathbf{r}) = 0$.

MONODROMY. Describe now the constructive way of getting the desired conformal transform. The single-valued inverse function $z(\zeta) \equiv \zeta^{-1}(z)$ is defined in the fundamental domain of ζ—the triangle ABC. The multivalued function $\phi(\zeta)$ is determined as follows:
– The function $\phi(\zeta)$ coincides with $z(\zeta)$ in the basic fundamental domain;
– In all other domains of the covering space ζ the function $\phi(\zeta)$ is analytically continued through the boundaries of these domains by means of fractional transformations consistent with the action of the group G.

Consider two basic contours P_1 and P_2 on ζ being the conformal images of the contours C_1 and C_2 (fig.3.1b). The function $\phi(z)$ ($z \neq \{z_1, z_2, \infty\}$) obeys the following transformations:

$$\phi\left[z \xrightarrow{C_1} z\right] \to \tilde{\phi}_1(z) = \frac{a_1\phi(z) + b_1}{c_1\phi(z) + d_1}; \quad \phi\left[z \xrightarrow{C_2} z\right] \to \tilde{\phi}_2(z) = \frac{a_2\phi(z) + b_2}{c_2\phi(z) + d_2}$$

(3.8)

where

$$\left(\begin{array}{cc} a_1 & b_1 \\ c_1 & d_1 \end{array} \right) = g_1; \qquad \left(\begin{array}{cc} a_2 & b_2 \\ c_2 & d_2 \end{array} \right) = g_2 \qquad (3.9)$$

are the matrices of basic substitutions of the group $G\{g_1, g_2\}$.

We assume $\zeta(z)$ to be a ratio of two fundamental solutions, $u_1(z)$, and, $u_2(z)$, of some second order differential equation with peculiar points $\{z_1 = (0, 0), z_2 = (0, 1), z_3 = (\infty)\}$. As it follows from the analytic theory of differential equations [8], the solutions $u_1(z)$ and $u_2(z)$ undergo the linear transformations when the variable z moves along the contours C_1 and C_2:

$$C_1 : \begin{pmatrix} \tilde{u}_1(z) \\ \tilde{u}_2(z) \end{pmatrix} = g_1 \begin{pmatrix} u_1(z) \\ u_2(z) \end{pmatrix} ; \quad C_2 : \begin{pmatrix} \tilde{u}_1(z) \\ \tilde{u}_2(z) \end{pmatrix} = g_2 \begin{pmatrix} u_1(z) \\ u_2(z) \end{pmatrix}$$
(3.10)

The problem of restoring the form of differential equation knowing the monodromy matrices g_1 and g_2 of the group G known as Riemann-Hilbert problem has an old history [8,9]. In our particular case we restrict ourselves with the well investigated groups Γ_2 (the free group) and $PSL(2,\mathbb{Z})$ (the modular group). The corresponding monodromy matrices are defined in (2.16) and (2.28). Thus, we have the following second-order differential equations:

$$z(z-1)\frac{d^2}{dz^2}u^{(f)}(z) + (2z-1)\frac{d}{dz}u^{(f)}(z) + \frac{1}{4}u^{(f)}(z) = 0$$
(3.11)

for the free group and

$$z(z-1)\frac{d^2}{dz^2}u^{(m)}(z) + \left(\frac{5}{3}z - 1\right)\frac{d}{dz}u^{(m)}(z) + \frac{1}{12}u^{(m)}(z) = 0$$
(3.12)

for the modular group.

The function which performs the conformal mapping of the upper halfplane Im$z > 0$ on the fundamental domain (the curvilinear triangle ABC) of the universal covering ζ now reads

$$\zeta(z) = \frac{u_1(z)}{u_2(z)}$$
(3.13)

where $u_1(z)$ and $u_2(z)$ are the basic solutions of (3.11) and (3.12) for Γ_2 and $PSL(2,\mathbb{Z})$ respectively.

As an example we give an explicit form of the complex potential $A(z)$ for the free group Γ_2. Substituting Eq.(3.2) for Eq.(3.13), we get

$$A(z) = \frac{d\zeta(z)}{dz} = \frac{1}{2(z-1)}\left(\frac{F_1(z)F_4(z)}{F_2^2(z)} - \frac{F_3(z)}{F_2(z)}\right)$$
(3.14)

where

$$F_1(z) = \int_1^{1/\sqrt{z}} \frac{d\kappa}{\sqrt{(1-\kappa^2)(1-z\kappa^2)}}; \quad F_2(z) = \int_0^1 \frac{d\kappa}{\sqrt{(1-\kappa^2)(1-z\kappa^2)}}$$

$$F_3(z) = \int_1^{1/\sqrt{z}} \sqrt{\frac{1-\kappa^2}{1-z\kappa^2}}d\kappa; \quad F_4(z) = \int_0^1 \sqrt{\frac{1-\kappa^2}{1-z\kappa^2}}d\kappa$$

The asymptotics of (3.14) is as follows

$$\frac{d\zeta(z)}{dz} \sim \begin{cases} \dfrac{1}{z} & z \to 0 \\[2mm] \dfrac{1}{z-1} & z \to 1 \end{cases}$$

(compare to the abelian case).

3.3. Random Walk on Double Punctured Plane and Conformal Field Theory

The geometrical construction described in the previous section is evidently related to the conformal field theory. In the most direct way this relation could be understood as follows. The ordinary differential equations Eq.(3.11) and Eq.(3.12) can be associated with equations on the four-point correlation function of some (still not defined) conformal field theory. The question remains whether it is always possible to adjust the central charge c of the corresponding Virasoro algebra and the conformal dimension Δ of the critical theory to the coefficients in equations like (3.11), (3.12). The question is positive and we show that on the example of the random walk on the double punctured plane with the monodromy of the free group.

We restrict ourselves to the "critical" case of infinite long trajectories, i.e. we suppose $L \to \infty$. In the field-theoretic language that means the consideration of the massless free field theory on z. Actually, the partition function of the selfintersecting random walk on z written in the field representation is generated by the scalar Hamiltonian $H = \frac{1}{2}(\nabla\varphi)^2 + m\varphi^2$ where the mass m functions as the "chemical potential" conjugated to the length of the path $(m \sim 1/L)$. Thus, for $L \to \infty$ we have $m_c = 0$ which corresponds to the critical point in conformal theory [4].

We introduce the conformal operator, $\varphi(z)$, on the complex plane z. The dimension, Δ, of this operator is defined from the conformal correlator

$$\langle \varphi(z)\varphi(z') \rangle \sim \frac{1}{|z - z'|^{2\Delta}} \tag{3.15}$$

Let us suppose $\varphi(z)$ to be a primary field, then the four-point correlation function $\langle \varphi(z_1)\varphi(z_2)\varphi(z_3)\varphi(z_4) \rangle$ satisfies the equation following from the conformal Ward identity [4,10,11]. In form of ordinary Riemann differential equation, Eq.(3.15) on the conformal correlator $\psi(z|z_1, z_2, z_3) =$

$\langle \varphi(z)\varphi(z_1)\varphi(z_2)\varphi(z_3) \rangle$ with the fixed points $\{z_1 = (0,0), z_2 = (1,0), z_3 = \infty\}$ reads [4,10]

$$\left\{ \frac{3}{2(2\Delta+1)}\frac{d^2}{dz^2} + \frac{1}{z}\frac{d}{dz} + \frac{1}{z-1}\frac{d}{dz} - \frac{\Delta}{z^2} - \frac{\Delta}{(z-1)^2} + \frac{2\Delta}{z(z-1)} \right\}$$

$$\psi(z|z_1, z_2, z_3) = 0$$

Performing the substitution

$$\psi(z|z_1, z_2, z_3) = [z(z-1)]^{-2\Delta} u(z)$$

we get the equation

$$z(z-1)u''(z) - \frac{2}{3}(1-4\Delta)(1-2z)u'(z) - \frac{2}{3}(2\Delta - 8\Delta^2)u(z) = 0 \quad (3.16)$$

which coincides with Eq.(3.11) for one single value of Δ

$$\Delta = -\frac{1}{8} \qquad (3.17)$$

The conformal properties of the stress-energy tensor, $T(z)$, are defined by the coefficients, L_n, in its Laurent expansion,

$$T(z) = \sum_{n=-\infty}^{\infty} \frac{L_n}{z^{n+2}}$$

These coefficients form the Virasoro algebra [4]

$$[L_n, L_m] = (n-m)L_{n+m} + \frac{1}{12}C(n^3 - n)\delta_{n+m,0}$$

where the parameter, C, is the central charge of the theory. Using the relation

$$C = \frac{2\Delta(5 - 8\Delta)}{(2\Delta + 1)}$$

established in [10] and Eq.(3.17) we obtain

$$C = -2 \qquad (3.18)$$

We find the following fact, mentioned by B. Duplantier, very intriguing. As he has pointed out, the value $\Delta = -\frac{1}{8}$ (Eq.(3.17)) coincides with the surface exponent (i.e. with the conformal dimension of the two point correlator near the surface) for the dense phase of the $O(n = 0)$ lattice model (or, what is the same, for the Potts model with $q = 0$) describing statistics of the so-called "Manhattan random walks" (known also as "dense polymers"—see the paper [12]). Recall that Potts model has been already mentioned in the Chapter 1 in connection with construction of algebraic knot invariants. It is hard to believe that such coincidence is occasional and we hope that the relation between these problems will be elucidated in the near future.

According to [12] the value $\Delta = -\frac{1}{8}$ belongs to the family of critical exponents

$$x_S \equiv x_{O(n=0)}^{surf} = \frac{S(S-2)}{8} \qquad (3.19)$$

where S is the number of fluctuating chains brought together in a bunch connecting the points z and z' on the complex plane (see fig.3.2 and [12] for more details). The critical behavior of the two-point correlation function for the "watermelon" configuration with S chains in the bunch has the following scaling form

$$G(|z - z'|, m_c) \sim \frac{1}{|z - z'|^{2x_S}} \qquad (3.20)$$

The case $S = 1$ corresponds to the conformal dimension of the primary fields $\varphi(z)$ considered above (see Eqs.(3.15)–(3.18)). For $S = 2$ Eq.(3.19) gives $x_S = 0$ and hence we could expect the logarithmic behavior of the correlation function (3.20). On the basis of the results of Chapter 2 we can directly establish such behavior for the contractible random loop on the plane with removed points.

The group Γ_2 has the metrics of the Cayley tree. Introducing the noneuclidean distance, η, on the Cayley tree, $C(\Gamma_2)$, between ends of the simplest watermelon configuration with $S = 2$ shown in fig.3.2, we have (see Eq.(2.20)):

$$G(\eta, N_1, N_2) \sim \frac{\eta^2}{N_1^{3/2} N_2^{3/2}} \exp\left\{ -\frac{\eta^2}{2}\left(\frac{1}{N_1} + \frac{1}{N_2} \right) \right\} \qquad (3.21)$$

where N_1 and N_2 are the lengths of the trajectories in the bunch. The mapping of the correlation function Eq.(3.21) of the contractible random walks on the universal covering $\mathfrak{F}$ onto the double punctured complex plane

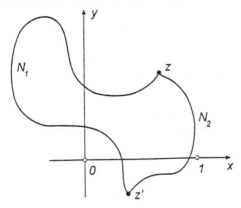

Fig. 3.2. Simplest "watermelon" configuration of the bunch of $S = 2$ chains having the trivial (i.e. contractible to the point) topological configuration with respect to peculiar points on z.

$\Re$ is given by the convolution (see also Ref.[13] and Section 4.1)

$$G\big(|z - z'|, N_1, N_2\big) \sim \int_0^\infty \frac{1}{\eta} \exp\left\{-\frac{|z - z'|^2}{\eta}\right\} G(\eta, N_1, N_2)d\eta \qquad (3.22)$$

The critical behavior of the correlation function (3.20) follows now from the Laplace transform:

$$G\big(|z - z'|, m_c\big) = \int G\big(|z - z'|, N_1, N_2\big)e^{m_c(N_1 + N_2)}dN_1 dN_2 \qquad (3.23)$$

Substituting Eq.(3.23) for Eqs.(3.21)-(3.22) we get

$$G\big(|z - z'|, m_c\big) \sim K_0\big(|z - z'|\sqrt{m_c}\big)\Big|_{m_c \to 0} \sim \ln|z - z'| \qquad (3.24)$$

Hence the scaling of the correlation functions defined in Eq.(3.24) is consistent with Eqs.(3.17)-(3.20), at least for the values $S = \{1, 2\}$.

The conformal invariance of the random walk [5,6] together with the geometrical interpretation of the monodromy properties of the four-point conformal correlator established above enable us to express the following assertion:

Statement 2 *The critical conformal field theory characterized by the values $C = -2$ and $\Delta = -\frac{1}{8}$ gives the field representation for the infinitely long random walk on the double punctured complex plane.*

The conformal field theory with $C = -2$ and $\Delta = -\frac{1}{8}$ has been recently studied as an example of the non-minimal model with the logarithmic-like singularities in the correlation functions. Thus, the behavior Eq.(3.24) could be considered as an additional confirmation of the statement expressed above.

With respect to the four-point correlation function, we could ask what happens with the gauge connection $A_j(z)$ if the argument z_j of the primary field $\psi(z_j)$ moves along the closed contour C around three punctures on the plane. From the general theory it is known that $A_j(z)$ can be written as

$$A_j(z) = \frac{2}{k} \sum_{i \neq j} \frac{R_i R_j}{z - z_i} \qquad (3.25)$$

where k is the level of the corresponding representation of the Kac-Moody algebra and R_i, R_j are the generators of representation of the primary fields $\psi(z_i)$, $\psi(z_j)$ in the given group [14].

The holonomy operator $\chi(C)$ associated with $A_j(z)$ reads

$$\chi(C) = P \exp \left(-\oint_C A_j(z) dz \right) \qquad (3.26)$$

It would be interesting to compare Eq.(3.14) (with one puncture at infinity) to Eq.(3.25). Besides we could also expect that Eq.(3.2) would allow us to rewrite the holonomy operator (3.26) as follows

$$\chi(C) = \exp \left(\zeta_{\text{in}} - \zeta_{\text{fin}} \right)$$

At this point we finish the brief discussion of the field-theoretical aspects of the geometrical approach presented above.

3.4. Statistics of Random Walks with Topological Constraints in 2D Lattice of Obstacles

The conformal methods can be applied to the problem of calculating the distribution function for random walks in regular lattices of topological obstacles on the complex plane $w = u + iv$. Let the elementary cell of the lattice be the equal-sided triangle with the side length c.

Introduce the distribution function $P(w_0, w, L|\text{hom})$ defining the probability of the fact that the trajectory of random walk starting at the point w_0 comes after "time" L to the point w and *all paths going from w_0 to w*

belong to the same homotopy class with respect to the lattice of obstacles. Formally we can write the diffusion equation

$$\frac{a}{4} \Delta_w P(w, L|\text{hom}) = \frac{\partial}{\partial L} P(w, L|\text{hom}) \qquad (3.27)$$

with initial and normalization conditions:

$$P(w, L = 0|\text{hom}) = \delta(z_0);$$

$$\sum_{\{\text{hom}\}} P(w_0, w, L|\text{hom}) = \frac{1}{\pi a L} \exp\left(-\frac{|w - w_0|^2}{aL}\right)$$

but it is very difficult even to formulate the problem analytically due to the absence of well-defined representation of the topological invariant. However the conformal methods can be used to find the asymptotic solution of Eq.(3.27) when $L \gg a$. In particular, we are interested in the probability to find the closed path of length L to be unentangled in the lattice of obstacles.

Following the general outline of Section 3.1 we construct the universal covering space $\zeta = \zeta(w)$; $\zeta = \xi + i\lambda$ for the multiconnected complex plane w. The complex plane ζ does not contain the branching point in any finite domain. Due to the conformal invariance of the Brownian motion, the new random process in the covering space will be again random but in the metric-dependent "new time".

The construction of the conformal transformation $\zeta(w)$ (explicitly described in [5]) can be performed in two steps—see fig.3.3:

1. First, by means of auxiliary reflection $w(z)$ we transfer the elementary cell of the w-plane to the upper halfplane of the $\text{Im}(z) > 0$ of the double punctured plane z. The function $w(z)$ is determined by the Christoffel-Schwarts integral

$$w(z) = \frac{c}{B\left(\frac{1}{3}, \frac{1}{3}\right)} \int_0^z \frac{d\tilde{z}}{\tilde{z}^{2/3} (1 - \tilde{z})^{2/3}} \qquad (3.28)$$

where $B\left(\frac{1}{3}, \frac{1}{3}\right)$ is the Beta-function. The correspondence of the branching points is as follows:

$$A(w = 0) \quad \rightarrow \quad \tilde{A}(z = 0)$$

$$B(w = c) \quad \rightarrow \quad \tilde{B}(z = 1)$$

$$C\left(w = c\, e^{-i\frac{\pi}{3}}\right) \quad \rightarrow \quad \tilde{C}(z = \infty)$$

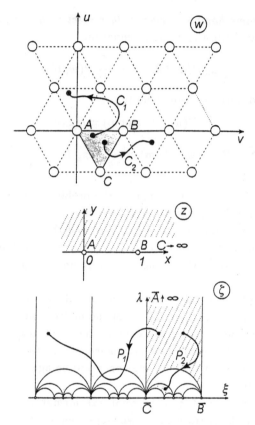

Fig. 3.3. Successive conformal transformations of the equal-sided triangle on the complex plane w to the fundamental domain $\bar{A}\bar{B}\bar{C}$ in the covering space ζ.

2. The construction of the universal covering ζ for the double punctured complex plane z is realized by means of automorphic functions. If the covering space is free of obstacles, the corresponding conformal transform should be as follows

$$-\frac{1}{\left(z'(\zeta)\right)^2}\{w(\zeta)\} = \frac{z^2 - z + 1}{2z^2(z-1)^2} \qquad (3.29)$$

where $\{z(\zeta)\}$ is the so-called Schwartz's derivative

$$\{w(\zeta)\} = \frac{z'''(\zeta)}{z'(\zeta)} - \frac{3}{2}\left(\frac{z''(\zeta)}{z'(\zeta)}\right)^2 ; \qquad z'(z) = \frac{dz}{d\zeta}$$

It is well known in the analytic theory of differential equations [8] that the solution of Eq.(3.29) can be represented as ratio of two fundamental solutions of some second order differential equation with two branching points, namely, of Eq.(3.11). The final answer reads

$$z(\zeta) \equiv k^2(\zeta) = \frac{\theta_2^4(0, e^{i\pi\zeta})}{\theta_3^4(0, e^{i\pi\zeta})} \tag{3.30}$$

where $\theta_2(0, \zeta)$ and $\theta_3(0, \zeta)$ are the elliptic Jacobi Theta-functions. We recall their definitions

$$\theta_2\left(\chi, e^{i\pi\zeta}\right) = 2e^{i\frac{\pi}{4}\zeta} \sum_{n=0}^{\infty} e^{i\pi\zeta n(n+1)} \cos(2n+1)\chi$$

$$\theta_3\left(\chi, e^{i\pi\zeta}\right) = 1 + 2\sum_{n=0}^{\infty} e^{i\pi\zeta n^2} \cos 2n\chi \tag{3.31}$$

The branching points $\tilde{A}$, $\tilde{B}$, $\tilde{C}$ have the images in the vertex points of zero-angled triangle lying in the upper halfplane of the plane ζ. We have from Eq.(3.30):

$$\tilde{A}(z = 0) \quad \rightarrow \quad \tilde{A}(\zeta = \infty)$$
$$\tilde{B}(z = 1) \quad \rightarrow \quad \tilde{B}(\zeta = 0)$$
$$\tilde{C}(z = \infty) \quad \rightarrow \quad \tilde{C}(\zeta = -1)$$

The halfplane $\mathrm{Im}(\zeta) > 0$ functions as a covering space for the plane w with the regular array of topological obstacles. It does not contain any branching point and consists of the infinite set of Riemann sheets, each of them having form of zero-angled triangle. These Riemann sheets correspond to the fibre bundle of w.

The conformal approach gives us a well defined nonabelian topological invariant for the problem—the difference between the initial and final points of the trajectory in the covering space (see Section 3.1). Thus, the diffusion equation for the distribution function $P(\zeta, L)$ in the covering space ζ with given initial point ζ_0 yields

$$\frac{a}{4} \frac{\partial^2}{\partial\zeta\partial\bar{\zeta}} P(\zeta, \zeta_0, L) = |w'(\zeta)|^2 \frac{\partial}{\partial L} P(\zeta, \zeta_0, L) \tag{3.32}$$

where we took into account that under the conformal transform the Laplace operator is transformed in the following way

$$\Delta_w = \left| \frac{d\zeta}{dw} \right|^2 \Delta_\zeta \qquad \text{and} \qquad \left| \frac{d\zeta}{dw} \right|^2 = \frac{1}{|w'(\zeta)|^2}$$

In particular, the value $P(\zeta = \zeta_0, \zeta_0, L)$ gives the probability for the path of length L to be unentangled (i.e. to be contractible to the point) in the lattice of obstacles.

The expression for the Jacobian $|w'(\zeta)|^2$ one can find using the properties of Jacobi Theta-functions [15]. Write $w'(\zeta) = w'(z)\, z'(\zeta)$, where

$$w'(z) = \frac{c}{B\left(\frac{1}{3}, \frac{1}{3}\right)} \frac{\theta_3^{16/3}}{\theta_2^{8/3} \theta_0^{8/3}}$$

and

$$z'(\zeta) = i\pi \frac{\theta_2^4 \theta_0^4}{\theta_3^4} ; \qquad i\frac{\pi}{4}\theta_0^4 = \frac{d}{d\zeta} \ln\left(\frac{\theta_2}{\theta_3}\right)$$

(we omit the arguments for compactness).

The identity

$$\theta_1'(0, e^{i\pi\zeta}) \equiv \left. \frac{d\theta_1(\chi, e^{i\pi\zeta})}{d\chi} \right|_{\chi=0} = \pi\theta_0(\chi, e^{i\pi\zeta})\, \theta_2(\chi, e^{i\pi\zeta})\, \theta_3(\chi, e^{i\pi\zeta})$$

enables us to get the final expression

$$|w'(\zeta)|^2 = c^2 h^2 \left| \theta_1'\left(0, e^{i\pi\zeta}\right) \right|^{8/3} \tag{3.33}$$

where

$$h = \frac{1}{\pi^{1/3} B\left(\frac{1}{3}, \frac{1}{3}\right)} \simeq 0.129$$

and

$$\theta_1(\chi, e^{i\pi\zeta}) = 2e^{i\frac{\pi}{4}\zeta} \sum_{n=0}^{\infty} (-1)^n e^{i\pi n(n+1)\zeta} \sin(2n+1)\chi \tag{3.34}$$

The 3D plot of the function $f(\xi, \lambda) = |w'(\xi + i\lambda)|^2$ is shown in fig.(3.4).

3.4.1. *Rare Lattice of Obstacles* $(a \ll L \ll c)$. *Correspondence to Edwards' approach*

Suppose that the random walk takes place in vicinity of some point, say, A in fig.3.3 far from other obstacles of the lattice. This condition is

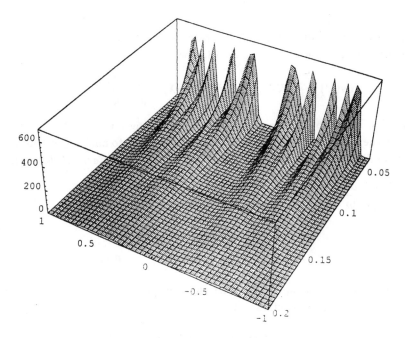

Fig. 3.4. Relief of Jacobian giving conformal mapping of plane with regular lattice of obstacles onto the universal covering.

fulfilled if the distance between the obstacles is much larger than the length of the trajectory: $c \gg L$. The conformal image of the point A lies in the infinity on the plane ζ. Thus, we can expand Eq.(3.33) for $\eta = \text{Im}(\zeta) \to \infty$. Using Eq.(3.34) we get

$$\theta'_1 \left(0, e^{i\pi(\xi + i\lambda)}\right)\bigg|_{\lambda \to \infty} = 2 \exp\left\{i\frac{\pi}{4}\xi\right\} \exp\left\{-\frac{\pi}{4}\lambda\right\} \qquad (3.35)$$

Substituting Eq.(3.35) into Eq.(3.33) we can rewrite the diffusion equation (3.27) in the following form

$$\frac{a}{4}\left(\frac{\partial^2}{\partial\xi^2} + \frac{\partial^2}{\partial\lambda^2}\right) P(\xi, \lambda, L) = 4c^2 h^2 \exp\left\{-\frac{2\pi}{3}\lambda\right\} \frac{\partial}{\partial L} P(\xi, \lambda, L) \qquad (3.36)$$

Eq.(3.36) can be obtained from Eq.(3.32) if the conformal transform of the punctured plane to the universal covering is realized via the logarithmic function

$$\zeta_{\text{single}} = -\frac{3}{\pi} \ln w + \text{const} \qquad (3.37)$$

The topological invariant in this case is the number of turns around the single obstacle A and therefore it coincides with the Gauss linking number considered in Section 1.2. Using Eq.(3.3) we get

$$\text{Inv}(C) = \oint_C \nabla \zeta_{\text{single}}(e_\xi d\xi + e_\lambda d\lambda) = \oint_C \frac{\xi d\lambda - \lambda d\xi}{\xi^2 + \lambda^2}$$

Rewriting Eq.(3.36) in polar coordinates and taking into account that one turn to the angle 2π corresponds to 6 successive turns to angles $\frac{\pi}{3}$ we obtain the final diffusion equation on z plane *with the single obstacle at the origin*

$$\frac{a}{4}\left\{\frac{1}{\rho}\frac{\partial}{\partial\rho}\left(\rho\frac{\partial}{\partial\rho}\right) + \frac{1}{\rho^2}\frac{\partial^2}{\partial\phi^2}\right\} P_0(\rho, \phi, L) = c^2 \frac{\partial}{\partial L} P_0(\rho, \phi, L) \qquad (3.38)$$

where

$$0 \le \phi \le 2\pi \quad \text{and} \quad P_0(\rho, \phi) = P_1(\rho, \phi + 2\pi) = \ldots = P_n(\rho, \phi + 2\pi n) \quad (3.39)$$

The solution of Eq.(3.38) under the condition (3.38) has the form of Eq.(1.11) discussed in Section 1.2.

3.4.2. *Dense Lattice of Obstacles* $(c^2 \ll La)$. *Random walk on Lobachevskii plane*

Return to Eq.(3.32) and perform the conformal transform of the upper halfplane $\text{Im}\zeta > 0$ to the interior of the unit circle on the complex plane τ in order to use the symmetry properties of the system. It is convenient to choose the following mapping of the vertices of the fundamental triangle $\bar{A}\bar{B}\bar{C}$

$$\bar{A}(\zeta = \infty) \quad \rightarrow \quad A'(\tilde{\zeta} = 1)$$
$$\bar{B}(\zeta = 0) \quad \rightarrow \quad B'(\tilde{\zeta} = e^{-i\frac{2\pi}{3}})$$
$$\bar{C}(\zeta = -1) \quad \rightarrow \quad C'(\tilde{\zeta} = e^{i\frac{2\pi}{3}})$$

The corresponding transform reads

$$\zeta(\tau) = e^{-i\frac{\pi}{3}}\frac{\tau - e^{i\frac{2\pi}{3}}}{\tau - 1} - 1 \qquad (3.40)$$

and the Jacobian $|w'(\tau)|^2$ takes the form

$$|w'(\tau)|^2 = \frac{3c^2 h^2}{|1 - \tau|^4}\left|\theta_1'\left(0, e^{i\pi\zeta(\tau)}\right)^{8/3}\right| \qquad (3.41)$$

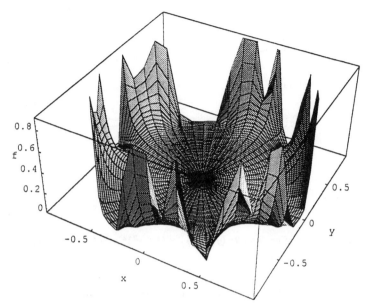

Fig. 3.5. Relief of the function $g(r, \psi)$—see explanations in the text.

In fig.(3.5) we plot the function $g(r, \psi) = \dfrac{1}{c^2}|w'(\tau)|^2$ where $\tau = r\,e^{i\psi}$.

The gain of such representation becomes clear if we average the function $g(r, \psi)$ with respect to ψ. The numerical calculations give us:

$$\lim_{r \to 1} \langle g(r, \psi)\rangle_\psi \equiv \lim_{r \to 1} \frac{1}{2\pi} \int_0^{2\pi} g(r, \psi)\,d\psi = \frac{\varpi}{(1 - r^2)^2} \tag{3.42}$$

where $\varpi \simeq 0.0309$ (see the fig.(3.6)).

Thus it is clear that for r rather close to 1 the diffusion is governed by the Laplacian on the surface of the constant negative curvature (the Lobachevskii plane). Representation of the Lobachevskii plane in the unit circle and in the upper halfplane (i.e. Poincare and Klein models) has been discussed in Section 2.4. Finally the diffusion equation (3.32) takes the following form:

$$\frac{\partial}{\partial N} P(r, \psi, N) = \mathcal{D}(1 - r^2)^2 \Delta_{r,\psi} P(r, \psi, N) \tag{3.43}$$

where $\mathcal{D} = \dfrac{a^2}{4\varpi c^2}$ is the "diffusion coefficient" in the Lobachevskii plane

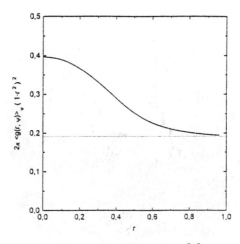

Fig. 3.6. Plot of product $2\pi \langle g(r, \psi)\rangle_\psi \times (1 - r^2)^2$ as a function of r.

and $N = L/a$ is the dimensionless chain length (i.e. effective number of steps).

Changing the variables $(r, \psi) \to (\mu, \psi)$, where $\mu = \ln \dfrac{1+r}{1-r}$, we get the unrestricted random walk on the 3-pseudosphere (see Eq.(2.84) of Section 2.4). The corresponding distribution function $P(\mu, N)$ reads

$$P(\mu, N) = \frac{e^{-\frac{N\mathcal{D}}{4}}}{4\pi\sqrt{2\pi(N\mathcal{D})^3}} \int_\mu^\infty \frac{\xi \exp\left(-\frac{\xi^2}{4N\mathcal{D}}\right)}{\sqrt{\cosh\xi - \cosh\mu}} d\xi \qquad (3.44)$$

The relation between the function $g(r, \psi)$ and the Lobachevskii plane metric becomes exact if we slightly modify the process of random walk in the lattice of obstacles and, therefore, the corresponding process on the universal covering. One can see that the surface $g(r, \psi)$ (fig.(3.5)) has the infinite set of local radial maxima. These maxima correspond to the centres of triangles on the original lattice of obstacles. Consider now the following process: the walker stays in the centre of some triangle, then randomly jumps with equal probabilities to any of the three centres of the neighboring triangles and so on... By choosing appropriate probabilities of staying and jumping, we can always adjust the average diffusion coefficient of such process to the diffusion coefficient $\mathcal{D}$ of the continuous random walk. To show that one can use the properties of the rational transformation

$$\tau \to \frac{\tau - \tau_0}{1 - \tau \, \bar{\tau}_0}$$

which maps the unit circle into itself, transferring the origin to some arbitrary internal point τ_0. Taking into account the fact that the point τ_0 is the centre of the lattice cell, we come after some algebra to Eq.(3.43). The corresponding calculations we leave for an exercise.

The physical meaning of the geodesics length on 3-pseudosphere, μ, is straightforward: μ is the length of so-called "primitive path" in the lattice of obstacles, i.e. length of the shortest path remaining after all topologically allowed contractions of the random trajectory in the lattice of obstacles. Hence, μ can be considered a nonabelian topological invariant, much more powerful than the Gauss linking number. This invariant is not complete except one point $\mu = 0$ where it precisely classifies the trajectories belonging to the trivial homotopic class.

3.5. Remarks and Conclusions

1. For double punctured plane $\mathcal{M}^2 = \mathbf{R}^2 - \{\mathbf{r}_1, \mathbf{r}_2\}$ with the generators $\{g_1, g_2\}$ attributed to the basis loops C_1, C_2 of the homotopy group $\pi_1 \left(\mathcal{M}^2 \right)$ (shown in fig.3.1a) it is possible to find an explicit analytic expression of the complete topological invariant *uniquely* characterizing any homotopy class of the paths on $\mathcal{M}^2$. For l-punctured plane $\mathcal{M}^2 = \mathbf{R}^2 - \{\mathbf{r}_1, \ldots, \mathbf{r}_l\}$ ($l \geq 2$) the problem of construction of nonabelian connections is related to the famous Riemann-Hilbert problem of restoring the form of differential equation with l peculiar points from the monodromy matrices $\{g_1, \ldots, g_l\}$. being deeply related to the Knizhnik-Zamolodchikov equations [11].

2. Comparing the distribution function of the primitive path μ to the distribution function of the knot complexity, η, defined previously (see Eq.(2.15)) we may conclude that both of these invariants have one and the same physical sense. This relation can be accounted for the following simple fact: the random walk in the covering space constructed for the lattice of obstacles is equivalent from topological point of view to the random walk on the Cayley tree (see fig.(3.3)). At the same time, as we have showed in Chapter 2, knot complexity is proportional to the length of the shortest noncontractible word written in terms of generators of the braid group B_n, i.e., is proportional to the length of geodesics in some space of constant

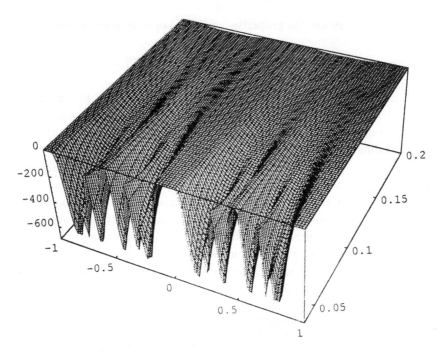

Fig. 3.7. Continuous analogue of ultrametric Cayley tree obtained by analytic function $-|w'(\zeta)|^2$ (see Eq.(3.33)).

negative curvature. The physical applications of the distribution functions for the primitive path are discussed in Chapter 4.

3. The problem considered here admits reformulation in spirit of spin-glass-like problems regarded in Chapter 1. Suppose there is a closed path of length L which is randomly dropped on the plane with the regular array of removed points. Let one point of the path be fixed. The question arises: what is the probability to find the random path in a specific topological state? Here, the topological state of a trajectory is a typical example of quenched disorder. To find the corresponding distribution function, the moments of the topological invariants have to be averaged over the Gaussian distribution (the measure for paths on the plane). The same speculations enable us to assume that the function $-|w'(\zeta)|^2$ (inverse of the Jacobian of conformal mapping)—see fig.3.7) has the sense of an ultrametric "phase space" discovered in [16], where each valley corresponds to some given topo-

logical state. The closer $\text{Im}\zeta$ is to the real axis, the higher are the barriers between neighboring valleys. Thus, all rather long $(La \gg c^2)$ random trajectories in such potential become "localized" in some heavily entangled state, in sense that the probability of spontaneous disentanglement of trajectory of length La is of order of $\exp\left(-\text{const}\dfrac{La}{c^2}\right)$. This analogy could be useful probably for the ordinary spin glass theory because of explicit expression of Parisi ultrametric potential in terms of analytic doubleperiodic functions.

References

1. A.M. Polyakov, Mod. Phys. Lett.(A), 3 (1988), 325
2. E. Witten, Comm. Math. Phys., 121 (1989), 351
3. S.K. Nechaev, A.Yu. Grosberg, A.M. Vershik, Preprint IHES/P/95/49; to appear in J. Phys. (A): Math. Gen. (1996)
4. A.A. Belavin, A.M. Polyakov, A.B. Zamolodchikov, Nucl. Phys.(B), 241 (1984), 333
5. S.K. Nechaev, J.Phys.(A): Math. Gen., 21 (1988), 3659
6. K. Ito, H.P. McKean, *Diffusion Processes and Their Sample Paths*, (Springer: Berlin, 1965)
7. D. Mumford, *Tata Lectutes on Theta I, II*, Progress in Mathematics, 28, 34, (Birkhauser: Boston, 1983)
8. V.V. Golubev, *Lectures on Analytic Theory of Differential Equations*, (GITTL: Moscow, 1950)
9. A.B. Venkov, Preprint LOMI P/2/86
10. Vl.S. Dotsenko, Nucl. Phys.(B) 235 [FS11] (1984), 54
11. V.G. Knizhnik, A.B. Zamolodchikov, Nucl. Phys. (B), 247 (1984), 83
12. B. Duplantier, F. David, J. Stat. Phys., 51 (1988), 327
13. S.K. Nechaev, A.N. Semenov, M.K. Koleva, Physica (A), 140 (1987), 506; L.B. Koralov, S.K. Nechaev, Ya.G. Sinai, Prob.Theor. Appl. (in Russian), (1993),
14. E.Guadarini, M.Martinelli, M.Mintchev, CERN-TH preprints 5419/89, 5420/89, 5479/89.
15. K Chandrassekharan, *Elliptic Functions*, (Springer: Berlin, 1985)
16. M. Mezard, G. Parisi, Virasoro, *Spin Glass Theory and Beyond*, (World Scientific: Singapore, 1987)

CHAPTER 4

PHYSICAL APPLICATIONS

4.1. Introduction: Polymer Language in Statistics of Entangled Chain-like Objects

Topological constraints essentially modify the physical properties of statistical systems consisting of chain-like objects of completely different nature. It should be said that topological problems are widely investigated in connection with quantum field and string theories, 2D-gravitation, statistics of vortexes in superconductors and world lines of anyons, quantum Hall effect, thermodynamic properties of entangled polymers etc. Modern methods of theoretical physics allow us to describe rather comprehensively the effects of nonabelian statistics on physical behavior for each particular referred system; however, in our opinion, the following general questions remain obscure:

(a) How does the changes in topological state of the system of entangled chain-like objects effect their physical properties?

(b) How can the knowledge accrued in statistical topology be applied to the construction of the Ginzburg-Landau-type theory of fluctuating entangled (nonabelian) chain-like objects?

In order to have representative and physically clear image for the system of fluctuating chains with the full range of nonabelian topological properties it appears quite natural to formulate general topological problems in terms of polymer physics. It allows us:

– To use a geometrically clear image of polymer with topological constraints as a model corresponding to the path integral formalism in the field theory;

– To advance in investigation of specific physical properties of biological and synthetical polymer systems where the topological constraints play a significant role.

It should be emphasized that the present Chapter is mainly concerned with the general problems of the topological properties of polymers represented as Brownian trajectories. The connection between topology and detailed chemical structure of polymers as well as the statistical properties of macromolecules caused by their specific chemical structure are beyond the scope of the present book.

For physicists the polymer objects are attractive due to many reasons. First of all, the adjoining of monomer units in chains essentially reduces all equilibrium and dynamic properties of the system under consideration. Moreover, due to that adjoining the behavior of polymers is determined by the space-time scales larger than for low-molecular-weight substances. This allows us to apply general theoretical methods such as perturbation theory, renormalization-group approach, conformal methods etc. to the investigation of polymer systems consisting of both ensemble of chains and a single macromolecule. Major progress in theoretical description of polymer systems is due to the combination of general methods of solid-state physics with the methods which take into account the chain-like structure of polymers. The chain-like structure of macromolecules causes the following peculiarities (see, for instance, [1]):

– The so-called "linear memory" (i.e. fixed position of each monomer unit along the chain);

– The low translational entropy (i.e. the restrictions on independent motion of monomer units due to the presence of bonds);

– Large space fluctuations (i.e. just a single macromolecule can be regarded as a statistical system with many degrees of freedom).

It should be emphasized that the above mentioned "linear memory" leads to the fact that different parts of polymer molecules fluctuating in space can not go one through another without the chain rupture. For the system of non-phantom closed chains this means that only those chain conformations are available which can be transformed continuously into one another:

which inevitably give rise to the problem of knot entropy determination (see Chapter 1 for details).

After these preliminary remarks it becomes clear that many general topological problems dealing with statistics of chain-like objects can be formulated easily in polymer language.

When analyzing various approaches to statistics of polymer chains with topological constraints, we should distinguish between two basic groups of works which could be called "microscopical" and "phenomenological". The former deal with those methods which allow to investigate exactly solvable simple basic models using quite rigorous methods, i.e., the methods elaborated in Chapters 1-3. On the other hand, phenomenological approaches do not contain rigorous mathematical description of topological constraints and are frequently based on the geometrically clear conjectures about the character of polymer chain fluctuations. This allows us to achieve significant results with respect to the properties of complex polymer systems without the complete mathematical analysis. Theories of such type do not claim high level of accuracy, but nowadays they remain most useful for investigation of physically important properties of polymers with the topological constraints [2,37]. Our further research is connected with the microscopical consideration of some well-known problems of statistics of polymers with topological constraints.

It should be emphasized that we do not concern ourselves with the problem of influence of topological constraints on dynamic properties of polymer chains.

4.2. Polymer Chain in 3D Array of Obstacles: Critical Exponents for Gyration Radius

The 3D-model "polymer chain in an array of obstacles" (PCAO) (see fig.4.1) can be defined as follows ([3,4,5,2,6]). Suppose a polymer chain of length $L = Na$ is placed between the edges of the simple cubic lattice with the spacing c, where N and a are the number of monomer units in the chain and the length of the unit correspondingly. We assume that the chain cannot cross ("pass through") any edges of the lattice. This model combines geometrical clarity of image with the possibility to investigate the influence of entanglements on equilibrium and dynamic properties of polymers exactly.

The PCAO-model can be considered as the basis for mean-field-like self-consistent approach to the major problem of entropy calculation of ensemble of strongly entangled fluctuating chains. Namely, choose the test chain, specify its topological state and assume that the lattice of obstacles models the effect of entanglements with the surrounding chains ("background"). Neglecting the fluctuations of the background and the topological constraints which the test chain produces for itself, we lose information about the correlations between the test chain and the background. However even in the simplest case we arrive at some nontrivial results concerning statistics of the test chain caused by topological interactions with the background. This means that for the investigation of properties of real polymer systems with topological constraints it is not enough to be able to calculate the statistical characteristics of chains in lattices of obstacles, but it is also necessary to be able to adjust any specific physical system to the unique lattice of obstacles, which is much more complicated task.

So, let us begin taking a closed polymer chain without volume interactions (i.e. a chain with selfintersections) in the trivial topological state with respect to the 3D lattice of obstacles. It means that the chain trajectory can be continuously contracted to the point (compare to the 2D case discussed in Chapter 3). It is clear that because of the obstacles, the macromolecule will adopt more compact conformation than the standard random walk without any topological constraints. Let us find the degree of such "compactification", using simple scaling estimates.

It is convenient to begin with the lattice realization of the problem. In this case the polymer chain can be represented as a closed N-step random walk on a cubic lattice with the length of elementary step a being equal to the spacing of the array of obstacles, c. The general case $a \neq c$ will be considered later.

On the basis of the results of Chapters 2 and 3 it is possible to make the following assertion:

Statement 3 *The random walk on a 3D-cubic lattice in the presence of the regular array of topological constraints produced by uncrossible strings on the dual lattice is equivalent to the free random walk on the graph—the*

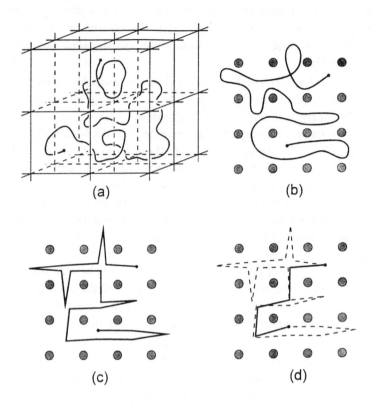

Fig. 4.1. Polymer chain in a array of obstacles: (a) 3D-case; (b) 2D-case; (c) trajectory coarsened up to scale of lattice of obstacles; (d) primitive path remaining after deleting of all double-folded chain parts.

Cayley tree with the branching number $z = 6$.

Now one can give the immediate answer to the question about the average space dimension $R(N) \equiv \sqrt{R^2(N)}$ of the closed unentangled N-step random walk ([5]). Divide the problem into two subsequent subparts:

1. First of all denote the geodesic length on the Cayley tree by k (i.e. k is the distance along the graph connecting initial and final points of some path). Call this geodesic length "the primitive path" (see fig(4.1d)) and consider N-step random walk on the tree starting and ending in the origin (the root point). Since this random walk is closed, it contains equal number of steps in the directions "from" and "to" the origin. Thus, the

average primitive path for the N-step closed path is $k \sim N^{1/2}$.

2. Now consider the embedding of the Cayley tree into the real 3D space. The spatial scale, R, corresponding to the k-step distance along the tree is equal to $R \sim ak^{1/2}$ because the k-step primitive path being embedded in the space is not affected by the topological constraints and has the Gaussian statistics.

Thus we have

$$R \sim aN^{1/4} \tag{4.1}$$

what means that typical conformation of closed polymer chain shown in fig.4.1 is strongly compressed with respect to the Gaussian conformation.

Let us turn now to exact derivation of the relation (4.1). It should be noticed that z-branching Cayley tree (called below as z-tree where $z = 2D$) functions as the universal covering considered in previous Chapters and can be regarded as a geometrical image of the free group Γ_∞ with an infinite number of generators. At the same time, $\Gamma_\infty / \mathbf{Z}^D = \Gamma_{z/2}$, where $\Gamma_{z/2}$ is the free group with z generators. Thus, the problem of the limit distribution of the random walks on z-tree follows from generalizations of the results obtained in Chapter 2. For instance, the probability $P(k, N)$ of the end-to-end distance for an N-step random walk on z-tree to obey the following recurrence relation

$$
\begin{aligned}
P(k, N+1) &= \tfrac{1}{z}P(k+1, N) + \tfrac{z-1}{z}P(k-1, N); & (k \geq 2) \\
P(k, N+1) &= \tfrac{1}{z}P(k+1, N) + P(k, N); & (k = 1) \\
P(k, N+1) &= \tfrac{1}{z}P(k+1, N); & (k = 0) \\
P(k, 0) &= \delta_{k,0}
\end{aligned}
\tag{4.2}
$$

compare to Eq.(2.17). Solving Eq.(4.2) as described in Chapter 2 we can easily find the conditional limit distribution for the function $P(k, m|N) = \dfrac{P(k, m)P(k, N-m)}{z(z-1)^{k-1}}$. Recall that $P(k, m|N)$ gives the conditional probability distribution of the fact that two subchains C_1 and C_2 of lengths m and $N - m$ have the common primitive path k under the condition that the composite chain $C_1 C_2$ of length N is closed and unentangled in regard to the obstacles;

$$P(k, m|N) \simeq \left(\frac{N}{2m(N-m)}\right)^{3/2} k^2 \exp\left(-\frac{k^2 N}{2m(N-m)}\right) \tag{4.3}$$

This equation enables us to get the following expressions for the mean length of the primitive path, $\langle k(m) \rangle$ of *closed unentangled N*-link chain divided into two parts of the lengths m and $N - m$ correspondingly

$$\langle k(m) \rangle = \sum_{k=0}^{N} k^2 P(k, m|N) \simeq \frac{2}{\sqrt{\pi}} \sqrt{\frac{2m(N-m)}{N}} \qquad (N \gg 1) \qquad (4.4)$$

The primitive path itself can be considered a random walk in a 3D space with restriction that any step of the primitive path should not be strictly opposite to the previous one. Therefore the mean-square distance in the space $\langle (\mathbf{r}_0 - \mathbf{r}_m)^2 \rangle$ between the ends of the primitive path of $k(m)$ steps is equal to

$$\langle (\mathbf{r}_0 - \mathbf{r}_m)^2 \rangle = \frac{z}{z-2} k a^2 \qquad (4.5)$$

where $\mathbf{r}_m$ is the radius-vector of a link with the number m and the boundary conditions are: $\mathbf{r}_N = \mathbf{r}_0 = 0$. The mean-square gyration radius, $\langle R_g^2 \rangle$ of N-step closed unentangled random walk in the regular lattice of obstacles reads

$$\begin{aligned} \langle R_g^2 \rangle &= \frac{1}{2N^2} \sum_{n \neq m} \langle (\mathbf{r}_n - \mathbf{r}_m)^2 \rangle = \frac{1}{2N} \sum_{m=1}^{N} \langle (\mathbf{r}_0 - \mathbf{r}_m)^2 \rangle \\ &= \frac{z}{z-2} \frac{\sqrt{2\pi}}{8} a^2 \sqrt{N} \end{aligned} \qquad (4.6)$$

This result should be compared to the mean-square gyration radius of the closed chain without any topological constraints,

$$\langle R_{g,0}^2 \rangle = \frac{1}{12} a^2 N$$

Arguments in support of this statement can be found in [7].

The problem of the random walk statistics in the lattice of obstacles with spacing $c > a$ (i.e. when the topological obstacles are situated not in every plaquet of the lattice) can also be mapped on the random walk on some graph representing the covering space. Nevertheless the topology of the graph in that case will be not as simple as one of the Cayley tree (see Chapter 2 and next Sections for details). However on the qualitative level the expression of the mean size of the unentangled closed loop in the regular lattice of obstacles can be obtained through the following scaling arguments. Represent a polymer chain as a sequence of "blobs" of size c, each of them containing $g \sim (c/a)^2$ monomer units. On scales smaller than c the entanglements are not essential and the walk is random without any constraints. On scales larger than c the conformation of the chain of blobs must be analogous to the conformation of the closed chain on the tree, i.e. the mean size of such chain of blobs should be of order $R \sim c(N/g)^{1/4}$, where N/g is the number of blobs in the macromolecule. Substituting the estimate for g we get for $c > a$

$$R \sim (ac)^{1/2} N^{\nu}; \qquad \nu = \frac{1}{4} \qquad (4.7)$$

The relation $R \sim N^{1/4}$ is reminiscent of the well-known expression for the dimension of randomly branched ideal macromolecule. The gyration radius of an ideal "lattice animal" containing N links is proportional to $N^{1/4}$. It means that both systems *belong to the same universality class*.

Now we turn to the mean-field calculation of the critical exponent ν of nonselfintersecting random walk in the regular lattice of obstacles [8]. Within the framework of Flory-type mean-field theory the nonequilibrium free energy, $F(R)$, of the polymer chain of size R with volume interactions can be written as follows

$$F(R) = F_{int}(R) + F_{el}(R) \qquad (4.8)$$

where $F_{int}(R)$ is the energy of the chain selfinteractions and $F_{el}(R)$ is the "elastic", (i.e. pure entropic) contribution to the total free energy of the system. Minimizing $F(R)$ with respect to R for fixed chain length, $L = Na$, we get the desired relation $R \sim N^{\nu}$.

Write the interacting part of the chain free energy written in the virial expansion

$$F_{int}(R) = V \left(B\rho^2 + C\rho^3 \right) \qquad (4.9)$$

where $V \sim R^d$ is the volume occupied by the chain in d-dimensional space; $\rho = \frac{N}{V}$ is the chain density; $B = b\frac{T-\theta}{\theta}$ and $C = \text{const} > 0$ are the two– and three– body interaction constants respectively. In the case $B > 0$ third virial coefficient contribution to Eq.(4.9) can be neglected [1].

The "elastic" part of the free energy $F_{el}(R)$ of an unentangled closed chain of size R and length Na in the lattice of obstacles can be estimated as follows

$$F_{el}(R) = \text{const} + \ln P(R, N) = \text{const} + \ln \int dk \, P(k, N) \, P(R, k) \quad (4.10)$$

where the distribution function $P(k, N)$ is the same as in Eq.(4.3) and $P(R, k)$ gives the probability for the primitive path of length k to have the space distance between the ends equal to R:

$$P(R, k) = \left(\frac{D}{2\pi a c k}\right)^{d/2} \exp\left(-\frac{DR^2}{2ack}\right) \tag{4.11}$$

Substituting Eq.(4.10) for Eqs.(4.3) and (4.11) we get the following estimate

$$F_{el}(R) = -\left(\frac{R^4}{a^2 c^2 N}\right)^{1/3} + o\left(R^{4/3}\right) \tag{4.12}$$

Equations (4.9) and (4.12) allow us to rewrite Eq.(4.8) in the form

$$F(R) \simeq B\frac{N^2}{R^D} - \left(\frac{R^4}{a^2 c^2 N}\right)^{1/3} \tag{4.13}$$

Minimization of Eq.(4.13) with respect to R for fixed N yields

$$R \sim B^{3/(4+3D)}(ac)^{2/(4+3D)}N^\nu; \qquad \nu = \frac{7}{4+3D} \tag{4.14}$$

The upper critical dimension for that system is $D_{cr} = 8$. For $D = 3$ Eq.(4.14) gives

$$R \sim N^{7/13} \tag{4.15}$$

It is interesting to compare Eq.(4.14) to the critical exponent ν_{an} of the lattice animal with excluded volume in the D–dimensional space, $\nu_{an} = \frac{3}{4+D}$, which gives $\nu_{an} = \frac{3}{7}$ for $D = 3$. The difference in exponents signifies that the unentangled ring with volume interactions and the nonselfintersecting "lattice animal" belong to *different universality classes* (despite in the absence of volume interactions they belong to the same class).

4.3. High Elasticity of Polymer Networks

A polymer network, also called *gel*, in the simplest case can be represented as a 3D–aggregate of polymer chains crosslinked by valent chemical bonds (see fig.4.2). The crosslinks can be produced, for instance, by irradiation of polymers in a solution or a melt.

Obviously, the physical behavior of gels depends strongly on the methods of their preparation. If the density of the crosslinks is low, the subchains between crosslinks are rather long and gel possesses the highelastic properties, i.e. it demonstrates extremely high elastic strains in the region of the nonlinear dependence of the strain on the stress. Construction of a molecular theory of nonlinear high elasticity of polymer chains presents one of the fundamental problems of statistical physics of macromolecules.

The classical theory of high elasticity was developed in the 1940s by a number of independent authors (see [9] for review). The main simplification of this theory consists in the assumption that the subchains obey the Gaussian statistics (i.e. the topological constraints are not taken into consideration) and the elasticity of a polymer network has a purely entropic nature. For uniaxial extenuation-compression of a dry (without a solvent) polymer network, the classical theory predicts the following dependence of stress, τ, per unit cross-section area of the sample in the initial state on the relative strain, $\lambda = l/l_0$ (l and l_0 are the dimensions in the direction of stretching axis after and before application of the stress respectively):

$$\tau = \frac{n_0 T}{V_0}(\lambda - \lambda^{-2}) \tag{4.16}$$

where T is the temperature, n_0 is the number of subchains in the sample and V_0 is its volume in the undeformed state. Eq.(4.16) agrees with the experimental data in the compression region ($\lambda < 1$) but for $\lambda > 1$ considerable deviations appear. These deviations are described by an empirical formula known as Mooney-Rivlin equation

$$\tau = \frac{n_0 T}{V_0}(\lambda - \lambda^{-2})(c_1 + c_2\lambda^{-1}) \tag{4.17}$$

where the constants c_1 and c_2 are in general of the same order of magnitude.

Construction of the microscopic base of Eq.(4.17) as well as of a set of other observed deviations from the classical theory of high elasticity is the subject of a large number of papers. Present day scientific thought finds the reason for these deviations in strong topological constraints between different subchains in the network. In the next subsection we review the molecular theory of high elasticity of polymer gels which has been recently developed on the basis of the model "polymer chain in an array of obstacles" (see the works [10,11]).

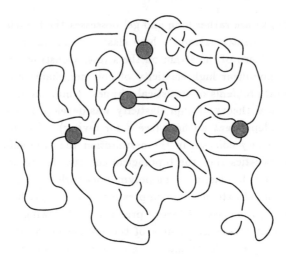

Fig. 4.2. Schematic representation of polymer gel

4.3.1. *Free Energy of Polymer Network*

Following Khokhlov and Ternovskii [10] let us proceed with the quantitative formulation of the problem. Assume that in the reference (undeformed) state each subchain of the polymer network is placed in a lattice of obstacles as shown in fig.4.1 and suppose for simplicity that all the subchains have the same length $L = Na$ (c is the spacing of lattice of obstacles). We consider the gel weakly crosslinked, $Na^2 \gg c^2$, i.e. each subchain occupies many cells of the lattice of obstacles. In a concentrated polymer system, such as polymer network, it is possible, as the first approximation, to disregard the volume interactions of links [12]. The individual chains are considered therefore statistically independent.

Introduce the coordinate system (x, y, z) connected to the lattice of obstacles and assume that the stress $(\lambda_x, \lambda_y, \lambda_z)$ is applied to the gel so that the sample volume remains unchanged: $\lambda_x = \lambda$; $\lambda_y = \lambda_z = \lambda^{-1/2}$. It is natural to expect the lattice of obstacles to deform affinely with the gel sample. Thus, in the deformed state the unit cell has the dimensions $\lambda_i c$ ($i = \{x, y, z\}$).

In the presence of the topological constraints the partition function of a subchain of length L with the fixed end points depends not only on the position of the ends (as in classical theories) but also on the topological state of the trajectory with respect to the lattice of obstacles. To clarify the point return to fig.4.1. Figure 4.1b shows the true microscopic trajectory of a subchain (in 2D case for simplicity), whereas fig.4.1c shows the same trajectory "coarsened" up to the scale c. Further coarsening can be achieved by eliminating all "double-folded" sections of the trajectory which are not entangled with the lattice of obstacles. Hereafter the trajectory in fig.4.1c is referred to as "coarsened" and the one shown in fig.4.1d as "primitive". Since the primitive path for a chain in the lattice of obstacles functions as a topological invariant, all conformations of the subchains with the same primitive path are topologically equivalent.

We show later that the partition function of the subchain numbered by α ($\alpha = 1, 2, \ldots, n_0$) depends only on the number $k_{\alpha i}$ of steps which the primitive path makes along the coordinate axes ($k_x = 3$ and $k_y = 2$ in fig.4.1b). Thus, $Z_\alpha = Z(N, \mathbf{k}_\alpha)$, $\mathbf{k}_\alpha = (k_{\alpha x}, k_{\alpha y}, k_{\alpha z})$ and the total partition function of the network reads:

$$Z = \prod_{\alpha=1}^{n_0} Z(N, \mathbf{k}_\alpha) \tag{4.18}$$

Denote by $n_0(\mathbf{k})$ the number of chains in the gel with primitive paths characterized by k_i and by $P(\mathbf{k}) = n_0(\mathbf{k})/n_0$ the corresponding distribution function which is not changed in course of gel deformation. The free energy, $\mathcal{F} = -T \ln Z$ of a gel sample can be rewritten using Eq.(4.18) as follows

$$\mathcal{F} = -n_0 T \sum_{k_i} P(\mathbf{k}) \ln Z(N, \mathbf{k}) \tag{4.19}$$

We proceed with calculation of $Z(N, \mathbf{k})$. Denote by Q_i the total number of steps of the coarsened trajectory along the ith coordinate axis. Some of these steps belong to the primitive path, other remain in the loop sections. In the deformed state the steps of the coarsened trajectory along the different coordinate axes do not generally form pairs since their lengths $\lambda_i c$ are different. Therefore the number of chain links along each axis is converged to these steps on average. Denote the corresponding quantity by G_i. Since the total number of subchain links is N, we should have in any conformation

$$\sum_i Q_i G_i = N \tag{4.20}$$

Now attribute weight p_i to each step along ith coordinate axis of the coarsened trajectory. Then the total coarsened trajectory carries the weight $\prod_i p_i^{Q_i}$. The calculation of p_i is presented below.

In consideration of the above said we arrive at the following expression for the partition function of one subchain in the lattice of obstacles

$$Z(N, \mathbf{k}) = \Gamma \sum{}' \prod_i p_i^{Q_i} Y(\mathbf{Q}, \mathbf{k}) \qquad (4.21)$$

where Γ is the total number of chain conformation, $Y(\mathbf{Q}, \mathbf{k})$ is the number of methods of realizing the coarsened trajectory with Q_i steps along the axes for a primitive path specified by parameters k_i; the prime in summation means that the sum is taken under the condition Eq.(4.20). Using the method of generating functions we get for Eq.(4.21)

$$Z(N, \mathbf{k}) = \frac{\Gamma}{2\pi i} \oint \frac{dx}{x^{N+1}} w(\mathbf{q}, \mathbf{k}) \qquad (4.22)$$

where $\mathbf{q}$ is a vector with components

$$q_i = p_i x^{G_i} \qquad (4.23)$$

and $w(\mathbf{q}, \mathbf{k})$ has the following expression

$$w(\mathbf{q}, \mathbf{k}) = \sum_{Q_i=0}^{\infty} \prod_i q_i^{Q_i} Y(\mathbf{Q}, \mathbf{k}) \qquad (4.24)$$

The contour integral in Eq.(4.22) is taken along the loop enclosing the point $x = 0$.

4.3.2. *Calculation of Generating Function*

Denote by 2σ the number of directions in which every next step of the coarsened trajectory in fig.4.1 can be made. We assign a separate number for each direction on the simple cubic lattice as shown in fig.4.3. We show the way of calculating sums over coarsened trajectories of definite classes with each trajectory entering with a weight $\prod_{u=1}^{2\sigma} q_u^{L_u}$ where $q_u = q_{u+\sigma}$ is the weight of a step in the direction u and L_u is the number of steps in that direction. For example, for 3D cubic lattice we have

$$
\begin{aligned}
q_1 = q_4 = q_x \qquad q_2 = q_5 = q_y \qquad q_3 = q_6 = q_z \\
L_1 + L_4 = Q_x \qquad L_2 + L_5 = Q_y \qquad L_3 + L_6 = Q_z
\end{aligned}
\qquad (4.25)
$$

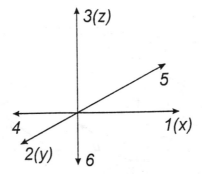

Fig. 4.3. Method of numbering directions for a 3D-cubic lattice of obstacles ($\sigma = 3$)

Let us introduce the following definitions:

1. $A_u(2m)$ is the sum over all closed trajectories (loops) not entangled with obstacles with the length of $2m$ which satisfy two conditions: a) the first step is made in the direction u, b) the trajectory returns to the starting point on the last step *for the firs time*. By definition $A_u(0) = 0$.

2. $B_u(2m)$ is the sum over all closed unentangled trajectories of length $2m$ returning to the initial point (not necessary for the first time) from all directions except one forbidden, u; $B_u(0) = 1$.

3. $C(2m)$ is the sum over all possible unentangled loops of length $2m$; $C(0) = 1$.

The quantities A_u, B_u, C satisfy the following Dyson-like relations:

$$A_u(2m) = q_u q_{u+\sigma} B_u(2m - 2)$$

$$B_u(2m) = \sum_{u \neq v} \sum_{n=0}^{m} A_v(2m - 2n) B_u(2n)$$

$$C(2m) = \sum_{\mu} \sum_{n=0}^{m} A_v(2m - 2n) C(2n) \tag{4.26}$$

Using the generating functions

$$\{A_u(y),\ B_u(u),\ C_u(y)\} = \sum_{m=0}^{\infty} \{A_u(2m),\ B_u(2m),\ C_u(2m)\}\, y^{2m} \tag{4.27}$$

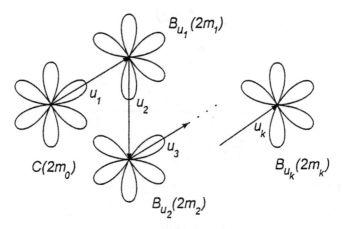

Fig. 4.4. Illustration of Eq.(4.29)

we get from Eqs.(4.26)

$$A_u(y) = y^2 q_u q_{u+\sigma} B_u(y)$$
$$B_u(y) = 1 + \left[A(y) - y^2 q_u q_{u+\sigma} B_u(y)\right] B_u(y) \qquad (4.28)$$
$$C(y) = 1 + A(y)C(y)$$

where we put

$$A(y) = \sum_u A_u(y)$$

Returning to Eq.(4.24) we can see that $w(\mathbf{q}, \mathbf{k})$ has a sense of a weighted sum over trajectories obtained by adding of all possible unentangled loops to the primitive path which is assumed to be specified by a sequence of directions $u_1, u_2, \ldots, u_k$. Using the quantities $C(2m)$ and $B_u(2m)$ introduced above, we can write

$$w(\mathbf{q}, \mathbf{k}) = q_{u_1} q_{u_2} \ldots q_{u_k} \sum_{m_0, m_1, \ldots, m_k = 0}^{\infty} C(2m_0) B_{u_1}(2m_1) \ldots B_{u_k}(2m_k)$$

$$= C(y = 1) \prod_u \left[q_u B_u(y = 1)\right]^{l_u}$$

$$(4.29)$$

where l_u is the number of primitive path steps in the u direction and $k_u = l_u + l_{u+\sigma}$. The equation (4.29) is illustrated in fig.4.4.

The quantities $C(y = 1)$ and $B_i(y = 1)$ can be easily determined from Eqs.(4.27)-(4.28). With $A(y = 1) = 1 - 2h$ we obtain

$$C(y = 1) = \frac{1}{2h}, \qquad B_i(y = 1) = \frac{1}{h + (h^2 + q_i^2)^{1/2}} \qquad (4.30)$$

where h is the non-negative solution of the equation

$$(\sigma - 1)h + \frac{1}{2} = \sum_{i=1}^{\sigma} (h^2 + q_i^2)^{1/2}$$

The integral in Eq.(4.22) can be calculated by the saddle-point method in case of our interests $Na^2 \gg c^2$. The result is as follows

$$\ln Z(N, \mathbf{k}) = \ln \Gamma - N \ln x + \sum_i k_i \ln \frac{q_i}{h + (h^2 + q_i^2)^{1/2}} \qquad (4.31)$$

where the value of x is determined from the condition that $\ln Z(N, \mathbf{k})$ is maximal with respect to x, i.e.

$$\frac{\partial \ln Z(N, \mathbf{k})}{\partial x} = 0 \qquad (4.32)$$

Equation (4.32) can be represented in the form

$$\sum_i G_i \overline{Q}_i = N$$

where

$$\overline{Q}_i = q_i \frac{\partial \ln Z(N, \mathbf{k})}{\partial q_i}$$

is the average number of steps of the coarsened trajectory along the i axis for all conformations with the given primitive path.

It should be noticed that number G_i of chain links per one step of coarsened trajectory (along the i axis) follows from the equation

$$\frac{\partial \ln Z(N, \mathbf{k})}{\partial G_i} = 0$$

which in turn gives

$$\frac{\partial p_i}{\partial G_i} + \ln x = 0$$

4.3.3. *Statistical Weight of Coarsened Trajectory and Averaging Over Primitive Paths*

To calculate the partition function $Z(N, \mathbf{k})$ explicitly we must find the statistical weights p_i for the steps of the coarsened trajectory characterized by the parameters Q_i. Here we present an approximate method of calculating p_i.

Let $\mathbf{r}_0, \mathbf{r}_1, \ldots, \mathbf{r}_Q$ be the coordinates of the segments of the roughened trajectory. The vector $\mathbf{r}_{s-1} - \mathbf{r}_s$ joints the same cells of the lattice of obstacles as the sth step of the coarsened trajectory. The number of microconformations for the given coarsened trajectory, $R = \Gamma \prod_i p_i^{Q_i}$, can be written pursuant to the following assumptions: (i) the chain part between points $\mathbf{r}_{s-1}$ and $\mathbf{r}_s$ contains exactly G_{i_s} links (i_s is the axis label along which the sth step is directed); (ii) all conformations of the chain part of G_{i_s} links correspond to the given coarsened trajectory. These assumptions allow us to write R as

$$R = \Gamma \int_{v_0} d\mathbf{r}_0 \int_{v_1} d\mathbf{r}_1 \ldots \int_{v_Q} d\mathbf{r}_Q \prod_{s=1}^{Q} \left(\frac{3}{2\pi a^2 G_{i_s}} \right)^{3/2} \exp\left[-\frac{3(\mathbf{r}_{s-1} - \mathbf{r}_s)^2}{2a^2 G_{i_s}} \right] \tag{4.33}$$

where $v_0, v_1, \ldots, v_Q$ are the volumes of the cells of the lattice of obstacles through which the coarsened trajectory of Q steps passes.

To simplify the result of integration over finite volumes $v_0, v_1, \ldots, v_Q$ let us assume that the integration can be extended to the whole space but with the appropriate Gaussian weight which provides the proper normalization:

$$\int_{v_s} d\mathbf{r}_s = \int d\mathbf{r}_s (2\pi)^{-3/2} \exp\left\{ -\sum_i \frac{[(\mathbf{r}_s)_i - (\mathbf{r}_{0s})_i]^2}{2\delta_i^2} \right\} \tag{4.34}$$

where $(\mathbf{r}_{0s})_i$ is the Cartesian coordinate of the centre of the corresponding cell of the lattice of obstacles and δ_i is its characteristic dimension along the i axis. Since the lattice of obstacles is supposed to be deformed affinely together with the gel sample, we can put

$$\delta_i = b\lambda_i \tag{4.35}$$

(obviously $b \sim c$; later on the ratio b/c is taken as adjusting parameter).

Substituting Eq.(4.33) for Eq.(4.34) and evaluating the remaining

Gaussian integrals we obtain

$$p = \text{const} \prod_j f\left(\frac{g_i}{\alpha\lambda_j^2}\right) \exp\left[-\frac{3\lambda_i^2}{2g_i}\right] \tag{4.36}$$

where

$$\alpha = 12\frac{b^2}{c^2}, \qquad g_i = G_i\frac{a^2}{c^2}, \qquad f(z) = \frac{1}{\sqrt{z} + \sqrt{z+1}} \tag{4.37}$$

The normalization constant should be determined from the condition that in the undeformed gel one has $p_i = \frac{1}{2\sigma}$.

To calculate the free energy of a polymer gel the logarithm of the partition function has to be averaged over the distribution of primitive paths $P(\mathbf{k})$ (see Eq.(4.31)). The partition function $P(\mathbf{k})$ depends in general on the method of the network preparation. Consider first the gel obtained by instantaneous crosslinking of chains in a *dry undeformed state*. In the case $Na^2 \gg c^2$ for the cubic lattice the function $P(\mathbf{k})$ is given by the solution of Eqs.(4.2) and has a sharp maximum of width $k^{1/2}$ near the point

$$k_i = \frac{k_0}{3} = \frac{2}{9}\frac{N}{G_0} \tag{4.38}$$

where G_0 is determined from Eq.(4.39) for the gel in the initial state.

As another example, consider networks obtained by the instantaneous crosslinking of polymer chains in a *concentrated solution*. In [10] it has been shown that in this case Eq.(4.38) can be replaced by the following one

$$k_i = \frac{\epsilon k_0}{3} \tag{4.39}$$

where the parameter ϵ reflects the degree of concentration of the polymer solution under the crosslinking conditions (obviously, $0 < \epsilon < 1$).

The averaging in Eq.(4.19) yields now

$$\mathcal{F} = -n_0 T \ln Z(N, \mathbf{k})\Big|_{k_i = \epsilon k_0/3} \tag{4.40}$$

In particular the reviewed theory has been applied to the description of the uniaxial extension-compression of a network prepared by instantaneous crosslinking. Numerous examples of application of this theory can be found in [10,11]. In physics of high elasticity of polymer networks the

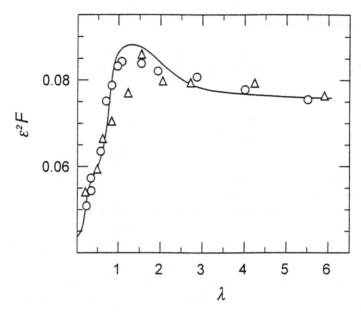

Fig. 4.5. Comparison of the theory (solid line) to the experimental data of Refs. [13] (circles) and [14] (triangles). The best agreement (correlation coefficient better than 0.99 in both cases) is reached by choosing the parameters $\alpha = 9$; $\varepsilon = 0.8$

dependence of the elastic stress $\tau = V_0^{-1} \partial \mathcal{F}/\partial \lambda$ on the relative strain is described as a rule in so-called Mooney-Rivlin coordinates (ξ, λ^{-1}), where $\xi = \tau/\tau_0$ with τ_0 taken from Eq.(4.16). The deviations of the plot $\xi(\lambda^{-1})$ from the straight line $\xi \equiv 1$ describe the corrections to the classical behavior. Comparison of the theory with the experimental results [13,14] is shown in fig.4.5. The best agreement is obtained for $\alpha = 9$; $\varepsilon = 0.8$. It should be noted that the correlation of the results of the theory with experimental data has turned out to be much higher than for other known theories of high elasticity (see [15]).

We believe that at present the molecular theory of elasticity based on the model "polymer chain in an array of obstacles" remains one of the most advanced theories of high elastic properties of regular gels. The generalization and refinement of the given analysis can be found in [11].

4.4. Collapsed Phase of Unknotted Polymer

In this section we show which predictions about the fractal structure of a strongly collapsed phase of unknotted ring polymer can be made using the concept of "polymer chain in array of obstacles".

Some attempts have been recently undertaken to consider theoretically the statistics of closed nonphantom fluctuating chains on the basis of nonabelian Chern-Simons topological field theory [16]. Although this idea seems to be very promising, we think that it is much too early to speak about any real progress. At the same time a qualitative description of thermodynamic properties as well as of fractal structure of nonphantom polymer with trivial topology can be developed in the region $\langle R_g^2 \rangle \ll \langle R_{g,0}^2 \rangle$ on the basis of the results obtained in Section 4.1 and in the previous Chapters within the framework of self-consistent analysis.

4.4.1. *"Crumpled Globule" Concept in Statistics of Strongly Collapsed Unknotted Polymer Loops*

Take closed nonselfintersecting polymer chain of length N in the trivial topological state [‡‡]. After a temperature decrease the formation of the collapsed globular structure becomes thermodynamically favourable [17]. Supposing that the globular state can be described in the virial expansion we introduce as usual two– and three–body interaction constants: $B = b\frac{T-\theta}{\theta} < 0$ and $C = \text{const} > 0$. But in addition to the standard volume interactions we would like to take into account the non-local topological constraints which obviously have a repulsive character. In this connection we express our main statement:

Statement 4 (Grosberg, Nechaev, Shakhnovich [18]) *The condition to form a trivial knot in a closed polymer changes significantly all thermodynamic properties of a macromolecule and leads to specific non-trivial fractal properties of a line representing the chain trajectory in a globule. We call such structure* **crumpled globule.**

We prove this statement consistently describing the given "crumpled" structure and showing its stability.

It is well-known that in a poor solvent there exists some critical chain length, g^*, depending on the temperature and energy of volume interac-

---

[‡‡]The fact that the closed chain cannot intersect itself causes two types of interactions: a) volume interactions which vanish for infinitely thin chains and b) topological constraints which remain even for chain of zero thickness.

tions, so that chains which have length bigger than g^* collapse. Taking long enough chain, we define these g^*-link parts as new block monomer units (crumples of minimal scale).

Consider now the part of a chain with several block monomers, i.e. crumples of smallest scale. This new part should again collapse in itself, i.e. should form the crumple of the next scale if other chain parts do not interfere with it. The chain of such new subblocks (crumples of new scale) collapses again and so on till the chain as a whole (see fig.4.6) forms the largest final crumple. Thus the procedure is completed when all initial links are united into one crumple of the largest scale. It should be noted that the line representing the chain trajectory obtained through the procedure described above resembles the 3D-analogue of the well known self-similar *Peano curve*.

It may seem that due to space fluctuations of the chain parts all that crumples could penetrate each others with the loops, destroying the self-similar scale-invariant structure described above. However it can be shown that if the chain length in a crumple of an arbitrary scale exceeds N_e then the crumples coming in contact do not mix with each other and remain segregated in space. Recall that N_e is the characteristic distance between neighbouring entanglements along the chain expressed in number of segments and, as a rule, the values of N_e lie in the range $30 \div 300$ [1].

Since the topological state of the chain part in each crumple is fixed and coincides with the state of the whole chain (which is trivial) this chain part can be regarded as an unknotted ring. Other chain parts (other crumples) function as effective lattice of obstacles surrounding the "test" ring— see fig.4.7. Using the results of the Section 4.1 (see Eq.(4.6)) we conclude that any M-link ring subchain without volume interactions not entangled with any of obstacles has the size

$$R^{(0)}(M) \sim aM^{1/4} \tag{4.41}$$

If $R^{(0)}$ is the size of an equilibrium chain part in the lattice of obstacles, the entropy loss for ring chain, S, as a function of its size, R, reaches its maximum for $R \simeq R^{(0)}$ (see Eq.4.12) and the chain swelling for values of R exceeding $R^{(0)}(M)$ is entropically unfavourable. At the same time in the presence of excluded volume the following obvious inequality must be fulfilled

$$R(M) \sim aM^{1/3}. \tag{4.42}$$

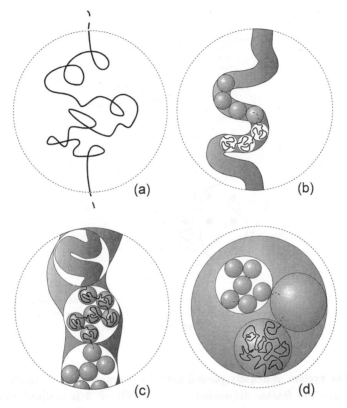

Fig. 4.6. (a)-(c) Subsequent stages of collapse; (d) Self-similar structure of crumpled globule segregated on all scales.

which follows from the fact that density of the chain in the globular phase $\rho \sim \dfrac{R^3}{N}$ is constant.

In connection with the obvious relation

$$R(M) > R^{(0)}(M) \tag{4.43}$$

we conclude that *the swelling of chains in crumples due to their mutual interpenetration with the loops does not result in the entropy gain and, therefore, does not occur in the system with finite density. It means that the size of crumple on each scale is of order of its size in dense packing state and the crumples are mutually segregated in space.* These questions are discussed in details in the work [18].

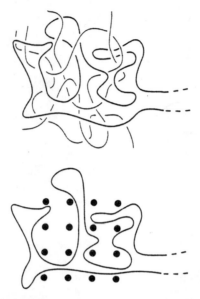

Fig. 4.7. (a) Part of the closed unknotted chain surrounded by other parts of the same chain; (b) Unentangled ring in lattice of obstacles. The obstacles replace the effect of topological constraints produced by other part of the same chain

The system of densely packed globulized crumples corresponds to the chain with the fractal dimension $D_f = 3$ ($D_f = 3$ is realized from the minimal scale, g^*, up to the whole globule size). The value g^* is of order

$$g^* = N_e(\rho a^3)^{-2}, \qquad (4.44)$$

where ρ is the globule density. This estimation was obtained in [18] using the following arguments: $g = (\rho a^3)^{-2}$ is the mean length of the chain part between two neighboring (along the chain) contacts with other parts; consequently $N_e g$ is the mean length of the chain part between topological contacts (entanglements). Of course, as to the phantom chains, Gaussian blobs of size g are strongly overlapped with others because pair contacts between monomers are screened (because of so-called θ-conditions [17]). However for nonphantom chains these pair contacts are topologically essential because chain crossings are prohibited for any value and sign of the virial coefficient.

The entropy loss connected with the crumpled state formation can be estimated as follows:

$$S \simeq -\frac{N}{g^*} \qquad (4.45)$$

Using Eq.(4.45) the corresponding crumpled globule density, ρ, can be obtained in the mean-field approximation via minimization of its free energy. The density of the crumpled state is less than that of usual equilibrium state what is connected with additional topological repulsive-type interactions between crumples:

$$\rho_{\text{crump}} = \frac{\rho_{eq}}{1 + \text{const}(a^6/CN_e)} < \rho_{\text{eq}} \qquad (4.46)$$

where ρ_{eq} is the density of the Lifshits' globule.

The experimental verification of the proposed self-similar fractal structure of the unknotted ring polymer in the collapsed phase meets some technical difficulties. One of the ways to justify the "crumpled globule" (CG) concept comes from its indirect manifestations in dynamic and static properties of different polymer systems. Among predictions and their justifications made on the basis of CG-concept we could mention: a) two-stage dynamics of collapse of the macromolecule after abrupt changing of the solvent quality, found in recent light scattering experiments by B. Chu and Q. Ying (Stony Brook) [19] b) one of possible explanations of the role of "introns" in the chromosome (DNA-molecule) compared to the DNA databank [20]; c) recent paper [21] where the authors claim the direct observation of the crumpled globule; and d) successive quantitative explanation of N-isopropylacrylamide gel collapse in pure water [22].

Let us mention also the very recent work [23] where on the basis of CG-concept the mechanism of "topologically driven" compatibility enhancement in mixtures of ring and linear chains has been proposed. Namely, it has been considered the thermodynamics of the bicomponent melts of linear chains A and unknotted unentangled closed chains B (trivial rings) within the framework of the usual Flory approach. It is shown in [23] that such melts exhibit a compatibility enhancement in comparison with the corresponding melts of linear chains. The reason for such effect is due to the topological entropy gain of mixing of A- and B-chains: if B-chains are unentangled since the A-chains play the role of a "diluent" softening the topological restrictions imposed on the B-chains. This result is in agreement with recent experimental findings of W. McKnight *et al* [24].

4.4.2. *Knot Formation Probability*

We can also utilize the CG-concept to estimate the trivial knot formation probability for dense phase of the polymer chain. Let us repeat that the main part of our modern knowledge about knot and link statistics has been obtained with the help of numerical simulations based on the exploiting of the algebraic knot invariants (Alexander, as a rule). Among the most important results we should mention the following ones:

– The probability of the chain self-knotting, $p(N)$, is determined as a function of chain length N under the random chain closure [25,26]. In the work [27] (see also the recent paper [28]) the simulation procedure was extended up to chains of order $N \simeq 2000$, where the exponential asymptotics of the type

$$p_0 \sim \exp(-N/N_0(T))$$

has been found for trivial knot formation probability for chains in good and θ-solvents. A statistical study of random knotting probability using the Vassiliev invariants has been undertaken in recent work [29].

– The knot formation probability p is investigated as a function of swelling ratio α ($\alpha < 1$) where $\alpha = \sqrt{\langle R_g^2 \rangle / \langle R_{g,0}^2 \rangle}$, $\langle R_g^2 \rangle$ is the mean-square gyration radius of the closed chain and $\langle R_{g,0}^2 \rangle = \frac{1}{12} N a^2$ is the same for unperturbed ($\alpha = 1$) chain—see fig.4.8 [25]. It has been shown that this probability decreases sharply when a coil contracts from swollen state with $\alpha > 1$ to the Gaussian one with $\alpha = 1$ [30,31] and especially when it collapses to the globular state [25,26].

– It has been established that in region $\alpha > 1$ the topological constraints are screened by volume interactions almost completely [30].

– It has been shown that two unentangled chains (of the same length) even without volume interactions in the coil state repulse each other as impenetrable spheres with radius of order $\sqrt{\langle R_{g,0}^2 \rangle}$ [25,32].

Return to fig.4.8, where the knot formation probability p is plotted as a function of swelling ratio, α, in the globular region ($\alpha < 1$). It can be seen that in compression region, especially for $\alpha < 0.6$ data of numerical experiment are absent. It is difficult to discriminate between different knots in strongly compressed regime because it is necessary to calculate Alexander polynomial for each generated closed contour. It takes of order $O(l^3)$ operations (l is the number of selfinteractions in the projection). This value becomes as larger as denser the system.

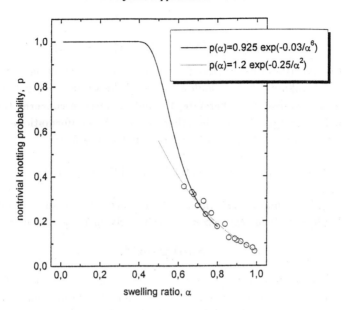

Fig. 4.8. Dependence of non-trivial knot formation probability, p on swelling parameter, α, in globular state. *Points*—data from Ref.[25]; *dashed line*—approximation in weak compression regime (Eq.(4.53)); *solid line*—approximation based on the concept of crumpled globule (Eq.(4.55)).

Let us present the theoretical estimations of the non-trivial knot formation probability $p(\alpha)$ in dense globular state ($\alpha < 0.6$) based on the CG-concept. The trivial knot formation probability under random linear chain closure, $q(\alpha) = 1 - p(\alpha)$, can be defined by the relation:

$$q(\alpha) = \frac{Z(\alpha)}{Z_0(\alpha)}, \qquad (4.47)$$

where $Z(\alpha)$ is the partition function of unknotted closed chain with volume interactions for fixed value of swelling parameter, α, and $Z_0(\alpha)$ is that of "shadow" chain without topological constraints but with the same volume interactions. Both partition functions can be estimated within the framework of the mean field theory. To do so, let us write down the classic Flory-type representation for the free energy of the chain with given α (in equations below we suppose for the temperature $T \equiv 1$):

$$
\begin{aligned}
F(\alpha) &= -\ln Z(\alpha) = F_{\text{int}}(\alpha) + F_{\text{el}}(\alpha) \\
F_0(\alpha) &= -\ln Z_0(\alpha) = F_{\text{int}}(\alpha) + F_{\text{el}}(\alpha)
\end{aligned}
\qquad (4.48)
$$

where
$$F_{el}(\alpha) = -S(\alpha), \qquad F_{el,0}(\alpha) = -S_0(\alpha) \qquad (4.49)$$

Here the contributions $F_{int}(\alpha)$ from the volume interactions to the free energies of unknotted and shadow chain of the same density (i.e. of the same α) are equivalent. Therefore, the only difference concerns the elastic part of free energy, F_{el}, or, in other words, the conformational entropy. Thus, the equation (4.47) can be represented in the form:

$$q(\alpha) = \exp\left(-F(\alpha) - F_0(\alpha)\right) = \exp\left(S(\alpha) - S_0(\alpha)\right) \qquad (4.50)$$

According to Fixmann's calculations [33] the entropy of phantom chain $S_0(\alpha)$ $(S_0(\alpha) = \ln Z_0(\alpha))$ in region $\alpha < 1$ can be written in the following form:

$$S_0(\alpha) \simeq -\alpha^{-2} \qquad (4.51)$$

Now we can write down the expression for the non-trivial knot formation probability $p(\alpha)$ depending only on the thermodynamic characteristic of polymer chain, $S(\alpha)$. Combining Eqs.(4.50) and (4.51) we get the following relation:

$$p(\alpha) = 1 - \exp\left(\alpha^{-2} + S(\alpha)\right) \qquad (4.52)$$

In the weak compression region $0.6 < \alpha \leq 1$ the probability of non-trivial knotting, $p(\alpha)$, can be estimated from the expression of the phantom ring entropy (Eq.(4.51)). The best fit of numerical data [25] gives us

$$p(\alpha) = 1 - A_1 \exp\left(-B_1\alpha^{-2}\right) \qquad (0.6 < \alpha \leq 1) \qquad (4.53)$$

where A_1 and B_1 are the numerical constant.

The nontrivial part of our problem is reduced to the estimation of the entropy of strongly contracted closed unknotted ring ($\alpha \ll 1$). Using Eqs.(4.44) and (4.45) and the definition of α we find

$$S(\alpha) \simeq -\frac{1}{N_e}\alpha^{-6} \qquad (4.54)$$

In the region of our interest ($\alpha < 0.6$) the α^{-2}-term in Eq.(4.52) can be neglected in comparison with α^{-6}. Therefore, we the final probability estimate has the form:

$$p(\alpha) = 1 - A_2 \exp\left(-\frac{1}{N_e}\alpha^{-6}\right) \qquad (\alpha < 0.6) \qquad (4.55)$$

where A_2 and N_e are the numerical constants.

The probabilities of the nontrivial knot formation, $p(\alpha)$, in weak and strong compression regions are shown in fig.4.8 by the dotted and solid lines respectively. The values of the constants are: $A_1 = 1.2$, $B_1 = 0.25$, $A_2 = 0.925$, $N_e = 34$; they are chosen by comparing Eqs.(4.53) and (4.55) with numerical data of Ref.[25].

4.4.3. *Quasi-Knot Concept in Collapsed Phase of Unknotted Polymers*

Speculations about the crumpled structure of strongly contracted closed polymer chains in the trivial topological state could be partially confirmed by the results of Chapters 1 and 2. The crucial question is: *why the crumples remain segregated in a weakly knotted topological state on all scales in course of chain fluctuations.* To clarify the point we begin by defining the topological state of a crumple, i.e. the unclosed part of the chain. Of course, mathematically strict definition of a knot can be formulated for closed (or infinite) contours exclusively. However the everyday experience tells us that even unclosed rope can be knotted. Thus, it seems attractive to construct a non-rigorous notion of a *quasiknot* for description of long linear chains with free ends.

Such ideas were expressed first in 1973 by I.M. Lifshits and A.Yu. Grosberg [34] for the globular state of the chain. The main conjecture was rather simple: in the globular state the distance between the ends of the chain is of order $R \sim aN^{1/3}$, being much smaller than the chain contour length $L \sim Na$. Therefore, the topological state of closed loop, consisting of the chain backbone and the straight end-to-end segment, might roughly characterize the topological state of the chain on the whole. The composite loop should be regarded as a quasiknot of the linear chain. The topological state of a quasiknot can be characterized by the knot complexity, η, introduced in Chapter 2. It should be noted that the quasiknot concept failed for Gaussian chains where the large space fluctuations of the end-to-end distance lead to the indefiniteness of the quasitopological state.

Our model of crumpled globule can be reformulated now in terms of quasiknots. Consider the ensemble of all closed loops of length L generated with the right measure in the globular phase. Let us extract from this ensemble the loops with $\eta(L) = 0$ and find the mean quasiknot complexity, $\langle \eta(l) \rangle$, of an arbitrary subpart of length l ($l/L = h = $ const; $0 < h < 1$) of the given loop. In the globular state the probability $\pi(\mathbf{r})$ to find the end of the chain of length L in some point $\mathbf{r}$ inside the globule of volume R^3 is of order $\pi(\mathbf{r}) \sim \dfrac{1}{R^3}$ being independent on $\mathbf{r}$ (this relation is valid when $La \gg R^2$). So, for the globular phase we could roughly suppose that the loops in the ensemble are generated with the uniform distribution. Thus our system satisfies the "brownian bridge" condition (see Chapter 2) and according to Conjecture 2 of Section 2.5 we can apply the following scale-invariant estimate for the averaged quasiknot complexity $\sqrt{\langle \eta^2(l) \rangle}$

$$\sqrt{\langle \eta^2(l) \rangle} \sim l^{1/2} = h^{1/2} L^{1/2} \qquad (4.56)$$

This value should be compared to averaged complexity $\sqrt{\langle \eta^2(l) \rangle}$ of the part of the same length l in the equilibrium globule created by an open chain of length L, i.e. without the brownian bridge condition

$$\sqrt{\langle \eta^2(l) \rangle} \sim l = hL \qquad (4.57)$$

Comparing Eqs.(4.56) and (4.57) we conclude that any part of an unknotted chain in the globular state is far less knotted than the same part of an open chain in the equilibrium globule, which supports our mean-field consideration presented above.

4.5. Ordering Phase Transition in Entangled "Directed Polymers"

Considerable number of works is devoted to analysis of the liquid-crystalline-type phase transitions in systems of long chain molecules (see for instance [35,36]). Apparently, at present the scope of problems dealing with the nematic-type ordering in polymers is one of the most studied branches of statistical physics of macromolecules. However, to the best of our knowledge all the existing theories do not take into account the effects caused by entanglements between the chains in such systems.

In the present Section we develop the simple mean-field theory of ordering phase transition in the system of entangled "directed polymers", i.e. in the "braid" of fluctuating polymer chains with fixed topology in an external field.

It should be emphasized from the very beginning that we do not claim to find a new kind of phase transitions or to describe a new class of real physical systems. We pursue only two goals:

- To construct the mean-field-like theory of fluctuating entangled chains utilizing the concept of *knot complexity* introduced in Chapter 2;

- To show how the presence of topological constraints modifies the standard nematic-like phase transition in bunch of "braided polymers".

4.5.1. *The model*

Consider an ensemble of M identical directed N-step chains with the segment length a embedded in 3D-space. In absence of any interactions, conformations of each directed polymer can be regarded as a particular "world line" (time-ordered trajectory) of a particle randomly jumping on the square lattice (x, y). Supposing the motion of particles discrete in time, let us label the "time slices" by index j; $(0 \leq j \leq N)$. Assume that the particle (walker) stays in the point $r_0 = (x_0, y_0)$ on the plane (x, y) at some initial moment in time, $j = 0$. At the next moment $j = 1$ the walker can randomly jump to any of four neighboring points $(x_0 \pm a_\perp, y_0)$, $(x_0, y_0 \pm a_\perp)$ where $a_\perp = a \sin \theta$ and θ is the angle between the chain segment and the "time" axis. Continuing this process recursively we obtain the "world line" of the particle representing the Markov chain with the local transitional probabilities $p(r_i | r_{i+1})$ where

$$
p(r_i | r_{i+1}) = \begin{cases}
\frac{1}{4} & (x_0, y_0) \to (x_0 + a_\perp, y_0) \\
\frac{1}{4} & (x_0, y_0) \to (x_0, y_0 + a_\perp) \\
\frac{1}{4} & (x_0, y_0) \to (x_0 - a_\perp, y_0) \\
\frac{1}{4} & (x_0, y_0) \to (x_0, y_0 - a_\perp)
\end{cases} \tag{4.58}
$$

In addition let us require that:

(i) at the time moment $j = N$ the walker returns to the starting point (x_0, y_0);

(ii) the trajectory of any particle on (x, y)-plane does not escape a disk of diameter $D \sim Na^2$.

The $(2 + 1)$–space-time– and 2D–representations of a particular configuration of mutually noninteracting world lines are shown in fig.4.9.

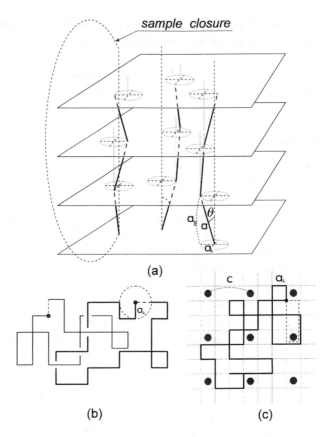

Fig. 4.9. Typical conformation of directed polymers in "space-time" representation (a), their (x, y)-projection (b) and corresponding random walk in the lattice of obstacles (c).

The partition function $\Omega_{\text{nonint}}(N, \theta, M)$ of M noninteracting identical N-step closed chains on the plane has the following simple expression

$$\Omega_{\text{nonint}}(N, \theta, M) = \left(\frac{4^N}{\pi N \sin^2 \theta} \right)^M \tag{4.59}$$

The corresponding free energy $F_{\text{nonint}} = -T \ln \Omega_{\text{nonint}}(N, \theta, M)$ is minimal when $\theta \to 0$.

Let us consider the interactions in the system of directed polymers.

TOPOLOGICAL INTERACTIONS AND ENTROPY. Imagine now the world

lines to be nonphantom. Because of the closure condition, each (x, y)-projection of such world line produces a knot diagram. A bunch of M mutually entangled world lines in $(2+1)$-dimensions can be regarded as a *braid-like* system of entangled directed polymers with fixed topology. As we already know, the topological state of linked paths can be roughly characterized by the "knot complexity", η, extracted from rigorously defined algebraic knot invariants (see Chapter 2). Our aim consists in calculating the entropy of fluctuating chains for fixed knot complexity η.

Formally our problem can be posed as follows. Consider the ensemble of all allowed chain conformations (with fixed ends) in the $(2+1)$-dimensional space-time. Suppose the chain topology is quenched in an arbitrary given state which does not change in course of thermal fluctuations. Due to the presence of topological constraints the entire phase space, Ω, splits into disconnected domains, $\omega\{\eta\}$, $(\omega \in \Omega)$ of "topologically similar" paths characterized by the knot complexity η. The entropy of chains with given η one can write down as follows

$$S\{\eta\} \equiv \ln \omega\{\eta\} = \ln \sum_{\{\Omega\}} \delta \left[\eta\{\mathbf{r}_0^1, \ldots, \mathbf{r}_0^M | \ldots | \mathbf{r}_N^1, \ldots, \mathbf{r}_N^M\} - \eta \right] \qquad (4.60)$$

where $\mathbf{r}_\alpha^j$ labels the position of jth segment of αth chain ($j \in [0, N]$, $\alpha \in [1, M]$).

As it has been already said, there is direct connection between the knot complexity η and the length of the shortest noncontractible (primitive) path μ in the lattice of obstacles. Let us utilize this relation constructing the simplest mean-field theory of topologically interacting world lines.

Conjecture 4 *Regard crossing of two **different** world lines in (x, y)-projection as a topological obstacle.*

The posed conjecture is of pure mean-field nature because we assume that different world lines interact topologically *through the fixed lattice of obstacles*. The advantage of the model under consideration consists in the fact that the selfintersections of world lines need not be taken into account. Actually, *each world line is phantom of itself* because of "time-ordering" condition (i.e. two distinct segments of one and the same trajectory belong to different time-ordered slices).

The partition function, $\omega(\mu, N, c, a_\perp)$, of the random walk of length $L = Na_\perp$ in the lattice of obstacles with the spacing, c, and the primitive path of length μ is given by the following equation

$$\omega(\mu, N, a_\perp, c) \simeq \left(\frac{c^2}{Na_\perp^2}\right)^{3/2} \frac{\mu}{c} \exp\left(\frac{Na_\perp^2}{c^2}\ln(2\sqrt{3}) + \frac{\mu}{2c}\ln 3 - \frac{\mu^2}{2Na_\perp^2}\right)$$
$$(4.61)$$

where the numerical coefficients correspond to the square lattice of obstacles and $a_\perp$ is the length of the segment projection to the plane (x, y).

In consideration of Conjecture 4 (i.e. supposing $\mu = \eta$), we can write down the expression for entropic (elastic) contribution to the free energy, F_{el}, of M fluctuating entangled world lines as a function of the link complexity, η:

$$F_{\text{el}}(\eta, N, M, D, a_\perp) = -M \ln \omega\left(\mu \equiv \eta, N, a_\perp c = \frac{D}{\sqrt{M}}\right) \simeq$$

$$-\frac{Na_\perp^2 M^2}{D^2}\ln(2\sqrt{3}) - \frac{\eta M^{3/2}}{2D}\ln 3 + \frac{M\eta^2}{2Na_\perp^2} - M \ln\left(\frac{\eta D^2}{(Na_\perp^2)^{3/2}M}\right)$$
$$(4.62)$$

where we have $T \equiv 1$ for the temperature and $c = D/\sqrt{M}$ for the average distance between the effective topological obstacles of the lattice shown in fig.4.9c.

2. INTERACTIONS WITH EXTERNAL FIELD. Suppose chains are placed in an external ordering field of arbitrary nature. It can be assumed for instance that the strain along the "time" axis is applied to the ends of all chains in the bunch. The potential energy of any chain segment in this external field reads

$$U(\mathbf{r}_j^\alpha) = g \cos \theta_j^\alpha \qquad (4.63)$$

where g is the interaction constant with the ordering field and $\theta_j^\alpha \equiv \theta$ according to the definition of the model (see fig.4.9a).

Total energy of the bunch of M directed polymers after applying of the strain can be written as follows

$$U(\mathbf{r}_0^1, \ldots, \mathbf{r}_N^M) \equiv U(N, M, \theta) = \sum_{\{j,\alpha\}} U(\mathbf{r}_j^\alpha) \qquad (4.64)$$

what gives in the mean-field approximation the simple expression for the interacting part of the free energy $F_{int} = -U(N, M, \theta)$

$$F_{\text{int}}(N, M, \theta) = -MNg\cos\theta \qquad (4.65)$$

Undoubtedly, our consideration is rather crude and does not take into account the steric intrachain interactions based on excluded volume effects. However we believe that being as simple as possible, the model describes the main features of physical behavior of entangled directed polymers under application of the strain.

In particular, we expect the phase transition from the disordered phase to the ordered one. In order to understand the nature of the transition let us begin by analysing the situation when chains in the bunch are strongly entangled. In this case some chains wind around others. Therefore, the case when all segments are parallel to the "time"-axis seems to be impracticable. On the other hand, the potential energy $U(\mathbf{r}_0^1, \ldots, \mathbf{r}_N^M)$ is maximal (for fixed g) when all segments are parallel to the "time"-axis. The competition between the entropic contribution from the chain fluctuations in fixed topological state and the potential energy of chain segments in external ordering field could lead to an ordering phase transition in the system under consideration. The less entangled the chains are the more favourable is the ordering transition. The corresponding phase diagram in the coordinates "link complexity" versus g (strength of interaction in Eq.(4.64)) is analysed at length in the next Section.

4.5.2. *Mean-field Theory of Phase Transition in System of Entangled Directed Polymers*

In the mean-field approximation the total free energy of the system, F, is the sum of "elastic", F_{el}, and "ordering", F_{int}, terms. Combining Eqs. (4.62) and (4.64) we obtain the following expression for the non-equilibrium free energy of the system of entangled directed random walks

$$
F(\theta) = -\frac{Na^2M^2\ln(2\sqrt{3})}{D^2}\sin^2\theta - \frac{\eta M^{3/2}\ln 3}{2D} + \frac{M\eta^2}{2Na^2}\sin^{-2}\theta
$$
$$
-M\ln\left(\frac{\eta D^2}{(Na^2)^{3/2}M}\sin^{-3}\theta\right) - gNM\cos\theta + \text{const}
$$

$$(4.66)$$

where $\sin^2\theta = w$ is the variational parameter changing in the region $w \in [\eta^2/(Na)^2, 1]$. In principle the free energy Eq.(4.66) should be minimized with respect to D (as well as to w) to reach the equilibrium density but we start with the simplified case supposing density to be constant.

Define the dimensionless density, ρ, and the relative length of the averaged primitive path (called below "relative link complexity"), τ, as

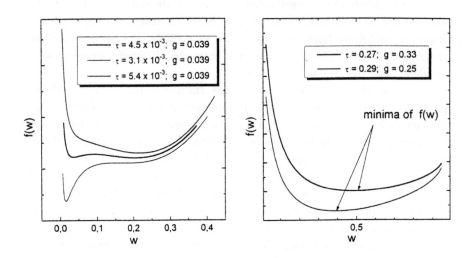

Fig. 4.10. Sample plots of the function $f(w)$ (Eq.(4.68)) for fixed values $\rho = 0.019$; $N = 3750$ and different relative link complexities, τ, and interaction constants, g.

follows

$$\rho = \frac{Ma^2}{D^2}; \qquad \tau = \frac{\eta}{Na} \quad (0 \leq \tau \leq 1) \tag{4.67}$$

The normalized free energy, $f(w)$, reads now

$$f(w) \equiv \frac{2}{NM} F(\sin^2 \theta = w) =$$

$$-\rho w \ln 12 + \frac{\tau^2}{w} + \frac{3}{N} \ln w - 2g\sqrt{1-w} + C(\rho, \tau, N) \tag{4.68}$$

where

$$\tau^2 \leq w \leq 1$$

and the function $C(\rho, \tau, N) = -\rho^{1/2}\tau \ln 3 + \frac{2}{N} \ln \rho$ does not depend on w.

The variable $w = \sin^2 \theta$ functions as an "order parameter" in our model. In the isotropic phase we have for the distribution function $\psi(\theta) = \frac{1}{\pi}$ ($\theta \in [0, \pi]$). Thus, $w_{\text{iso}} = \int_0^\pi w(\theta) \, \psi(\theta) \, d\theta = \frac{1}{2}$. Let us assume that

– for $w < \frac{1}{2}$ the chains are in the ordered (nematic-like) phase;

– for $w \geq \frac{1}{2}$ the chains are in the disordered phase. Note that actually the values of the order parameter w greater than $1/2$ correspond to the ordering in the layers normal to z-axis, but in the framework of the model we discuss the transition between two phases only—ordered (nematic-like) and disordered ones.

We determine the phase transition curve through comparison of the minimal values of the free energy $f(w_{min})$ on the interval $[\tau^2, w_{iso}]$ to the free energy $f(w_{iso} = \frac{1}{2})$ of disordered (by convention) phase.

It can be seen that the *first-order* phase transition could appear for values $g \leq \rho \ln 12$ exclusively. The condition for the first-order transition is

$$f(w = w_{min}^{(1)}) = f(w = w_{min}^{(2)}) \tag{4.69}$$

where $\left\{ w_{min}^{(1)}, w_{min}^{(2)} \right\}$ are the minima of the function $f(w)$ (Eq.(4.68)) on the interval $w \in \left[\tau^2, \frac{1}{2} \right]$. The corresponding sample plots of the function $f(w)$ for the fixed values $\rho = 0.019$; $N = 3750$ are shown in fig.(4.10a).

The *second-order* phase transition appears when the free energy minimum reaches the upper boundary $w = \frac{1}{2}$ of allowed values of w. The transition point in this case is determined by the following equation

$$w_{min} = \frac{1}{2} \tag{4.70}$$

where w_{min} is the single minimum of the function $f(w)$ (see (fig.4.10b)).

The complete phase diagram in the coordinates (τ, g) is presented in fig.4.11, where the border of the transition from the disordered to the ordered phase is drawn for particular parameters: $\{\rho = 0.019; N = 3750\}$ and consists of the following parts:

(a) parts of first-order transition curve AB with the segment BC shown by solid line;

(b) part of second-order transition curve CD shown by broken line.

The part of the phase diagram in between of the first-order and second-order transition curves corresponds to the "partially ordered" state with the order parameter $w_{min} < w < w_{iso}$.

The shape of the transition curves is hardly effected by change of parameters ρ and N. Moreover obtained phase diagram supports also our

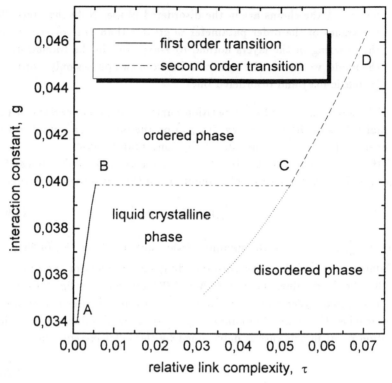

Fig. 4.11. Phase diagram of ordering transition in entangled bunch of directed random walks.

conjecture that weakly entangled directed polymers ($\eta \to 0$) can be ordered much easier than strongly entangled chains of the same lengths.

Thus, we have constructed the simple mean-field theory of the ordering transition in the system of entangled directed random walks in the broad region of the values of the "link complexity". We have also showed that the order of the phase transition varies for "weakly" and "strongly" entangled chains.

The ideas expressed in the present Section could be developed in the following directions:

- To consider the influence of the steric interchain segment-segment interactions on nematic-type ordering in the bunch of entangled directed polymers.

- To investigate the influence of topological constraints on the smectic-type ordering in the layers parallel to the (x, y)-plane.

- To extend the proposed theory beyond the mean-field approximation for analyzing the influence of global topological constraints on local correlation functions of chain segments.

4.6. Remarks and Conclusions

Let us summarize briefly our main findings in regard of the physical problems discussed above and represented in Tab.4.1.

First of all we would like to express the conjecture (see also [37]) concerning the possibility of reformulation of some topological problems for strongly collapsed chains (see Section 4.4) in terms of integration over the set of trajectories with fixed fractal dimension but without any topological constraints.

Conjecture 5 *We have argued that in ensemble of strongly contracted unknotted chains (paths) most of them have the fractal dimension $D_f = 3$;*

We believe that almost all paths in the ensemble of lines with fractal dimension $D_f = 3$ are topologically isomorphic to simple enough (i.e. close to the trivial) knot.

Let us remind, that the problem of the calculation of the partition function for closed polymer chain with topological constraints can be written as an integral over the set Ω of closed paths with fixed value of topological invariant (see Chapter 1):

$$Z = \int_{\Omega} D_w\{r\}e^{-H} = \int \ldots \int D_w\{r\}e^{-H}\delta[I - I_0], \qquad (4.71)$$

where $D_w\{r\}$ means integration with the usual Wiener measure and $\delta[I-I_0]$ cuts the paths with fixed value of topological invariant (I_0 corresponding to the trivial knots).

If our conjecture is true, then the integration over Ω in Eq.(4.71) for the chains in the globular phase (i.e. when $La \gg R^2$) can be replaced by the integration over all paths without any topological constraints, but with special new measure, $D_f\{r\}$:

$$Z = \int \ldots \int D_f\{r\}e^{-H} \qquad (4.72)$$

model invariant	Polymer chain near single obstacle	Knotting probability of polymer chain	Network high-elasticity	Coil-globule transition in trivial knot	Ordering in entangled bunch of directed chains
Gauss lin-king number	Analytic, Edwards, 1967	✕	Analytic, Tanaka, 1984 Edwards and Iwata, 1989	✕	✕
Algebraic polynomials	?	Numeric (Monte-Carlo), Vologodskii et al , 1974 Analytic, Grosberg and Nechaev, 1992	?	Numeric, Vologodskii et al , 1975 Koniaris and Muthukumar, 1991	?
Primitive path (a), "knot complexity" (b) and quasi-knot concept (c)	✕	Analytic (c), Grosberg and Nechaev, 1993 Analytic (b), Nechaev, Grosberg and Vershik, 1995	Analytic (a), Marrucci, 1987 Analytic (a), Khokhlov and Ternovskii, 1986-1989 Analytic (c), Grosberg and Nechaev, 1990	Analytic (c), Grosberg, Nechaev and Shakhnovich, 1989	Analytic (b), Nechaev, 1994

Table 4.1. Summary of physical problems dealing with statistics of entangled chain-like molecules and methods of their solutions.

The usual Wiener measure $D_w\{r\}$ is concentrated on trajectories with the fractal dimension $D_f = 2$. Instead of that, the measure $D_f\{r\}$ with the fractal dimension $D_f = 3$ for description of statistics of unknotted rings should be used.

References

1. A.Yu. Grosberg, A.R. Khokhlov, *Statistical Physics of Macromolecules* (AIP Press: New York, 1994)
2. S.K. Nechaev, Int. J. Mod. Phys.(B), 4 (1990), 1809
3. E. Helfand, D.S. Pearson, J. Chem. Phys., 79 (1983), 2054
4. M. Rubinstein, E. Helfand, J. Chem. Phys., 82 (1985), 2477
5. A.R. Khokhlov, S.K. Nechaev, Phys. Lett., 112A (1985), 156
6. M.K. Koleva, S.K. Nechaev, A.N. Semenov, Physica (A), 140 (1987), 506
7. L. Koralov, S.K. Nechaev, Ya.G. Sinai, Prob. Theory Appl., 38 (1993), 331
8. A.M. Gutin, A.Yu. Grosberg, E.I. Shakhnovich, Macromolecules, 26 (1993), 1293
9. L. Treloar, *The Physics of Rubber Elasticity*, 3rd ed., (Clarendon Press: Oxford, 1975)
10. A.R. Khokhlov, F.F. Ternovskii, Sov. Phys. JETP, 63 (1986), 728
11. A.R. Khokhlov, F.F. Ternovskii, E. Zheligovskaya, Physica (A), 163 (1990), 747

12. P.G. de Gennes, *Scaling Concepts in Polymer Physics* (Cornell Univ. Press: Ithaca, 1979)

13. H. Pak, J. Flory, J. Polym. Sci.: Polym. Phys.Ed., 17 (1979), 531

14. R.S. Rivlin, D.W. Saunders, Phil. Trans. Roy. Soc. (A), 243 (1951), 25

15. M. Gottlieb, R.J. Gaylord, Polymer, 24 (1983), 1644

16. A. Kholodenko, private communication

17. I.M. Lifshits, A.Yu. Grosberg, A.R. Khokhlov, Rev. Mod. Phys., 50 (1978), 683

18. A.Yu. Grosberg, S.K. Nechaev, E.I. Shakhnovich, J. de Physique, J. Phys. (Paris), 49 (1988), 2095

19. B. Chu, Q. Ying, A. Grosberg, Macromolecules, 28 (1995), 180

20. A. Grosberg, Y. Rabin, S. Havlin, A. Neer, Europhysics letters, v. 23, p. 373-378, 1993; A. Grosberg, Y. Rabin, S. Havlin, A. Neer Biofizika, v. 38, p. 75-83, 1993

21. J. Ma, J.E. Straub, E.I. Shakhnovich, J. Chem. Phys., 103 (1995), 2615

22. A.Yu. Grosberg, S.K. Nechaev, Macromolecules, 24 (1991), 2789

23. A.R Khokhlov, S.K. Nechaev, J. de Physique II (France), to appear

24. W.L. Nachlis, R.P. Kambour, W.J. McKnight, Abstracts of the 35th IUPAC Congress; vol. 2, p. 796; W.L. Nachlis, R.P. Kambour, W.J. McKnight, Polymer, 39 (1994), 3643

25. A.V. Vologodskii, M.D. Frank-Kamenetskii, Usp. Fiz. Nauk, 134 (1981), 641 (in Russian); A.V. Vologodskii, A.V. Lukashin, M.D. Frank-Kamenetskii, V.V. Anshelevich, Zh. Exp. Teor. Fiz., 66 (1974), 2153; A.V. Vologodskii, A.V. Lukashin, M.D. Frank-Kamenetskii, Zh. Exp. Teor. Fiz., 67 (1974), 1875; M.D. Frank-Kamenetskii, A.V. Lukashin, A.V. Vologodskii, Nature, 258 (1975), 398

26. J.P. Mishels, F.W. Wiegel, Proc. Roy. Soc. (A), 403 (1986), 269

27. K. Koniaris, M. Muthukumar, Phys. Rev. Lett., 66 (1991), 2211

28. K. Tsurusaki, T. Deguchi, J. Phys. Soc. Jap., 64 (1995), 1506

29. T. Deguchi, K. Tsurusaki, J. Knot Theory Ramific., 3 (1994), 321

30. K.V. Klenin, A.V. Vologodskii, V.V. Anshelevich, A.M. Dykhne, J. Biomol. Struct.Dyn.,5 (1988), 1173

31. E.J. van Rensburg, S.G. Whittington, J.Phys.(A):Math.Gen., 23 (1990), 3573

32. M. le Bret, Biopolymers, 19 (1980), 619

33. M. Fixmann, J.Chem.Phys., 36 (1962), 306

34. I.M. Lifshits, A.Yu. Grosberg, Zh. Exp. Teor. Fiz., 65 (1973), 2399

35. A. R. Khokhlov, A. N. Semenov, Physica (A), 108 (1981), 546; 112 (1982), 605

36. J. V. Seilinger, R. F. Bruinsma, Phys. Rev. (A), 43 (1991), 2910, 2922

37. A.Yu. Grosberg, S.K. Nechaev, Adv. Polym. Sci., 106 (1993), 1

Appendix A

LIMIT THEOREM FOR CONDITIONAL DISTRIBUTION OF PRODUCTS OF INDEPENDENT UNIMODULAR 2×2 MATRICES

The analysis of the asymptotic behavior of distributions of independent unimodular 2×2 matrices is a well-known subject in the probability theory. The classical paper by Fürstenberg [1] states, in particular, that under some natural conditions, the typical products of matrices grow exponentially.

Here we study (see also [2]) the asymptotical behavior of the *conditional* distribution over the matrix elements, supposing that the product as a whole belongs to a compact subset of the group $SL(2,\mathbf{R})$. To be more specific, let us assume that the probability distribution $\mathcal{P}$ on the group $SL(2,\mathbf{R})$ is defined and has the following properties:

(i) $\mathcal{P}$ is concentrated on a compact subset $T \subset SL(2,\mathbf{R})$;
(ii) $\mathcal{P}$ has the density $p(g)$, i.e.

$$\mathcal{P}(H) = \int_H p(g)dg, \qquad H \subset SL(2,\mathbf{R})$$

Consider the (left-hand) products $g^N = g_N g_{N-1} \ldots g_1$ where all $g_i \in SL(2,\mathbf{R})$ are independent and identically distributed with $\mathcal{P}$. Here we follow the technique of the so-called "large deviations" developed by Fürstenberg and Tutubalin (see [1,3]) and use special coordinates on the group $SL(2,\mathbf{R})$. Namely, each matrix

$$g = \left(\begin{array}{cc} a & b \\ c & d \end{array} \right)$$

we represent in the form

$$g = R_\alpha D_\lambda R_\beta \tag{A.1}$$

where $\{R_\alpha, R_\beta\} \in SO(2)$; $SO(2)$ is the abelian group of matrices

$$R_\varphi = \begin{pmatrix} \cos\varphi & \sin\varphi \\ -\sin\varphi & \cos\varphi \end{pmatrix}; \quad -\pi \le \varphi \le \pi;$$

and D_λ is the diagonal matrix

$$D_\lambda = \begin{pmatrix} \lambda & 0 \\ 0 & \lambda^{-1} \end{pmatrix}$$

with $\lambda \ge 1$.

The representation defined in the Eq.(A.1) is not-unique since

$$g = [-R_\alpha] D_\lambda [-R_\beta] \equiv R_{\alpha+\pi} D_\lambda R_{\beta+\pi}$$

It is convenient to pass to the unit tangle bundle over the group $PSL(2, \mathbf{R}) = SL(2, \mathbf{R})/(e, -e)$ (e is the unit of the group). The elements $h \in PSL(2, \mathbf{R})$ are pairs $(g, -g)$. Since $-g$ is represented by Eq.(A.1) with $\alpha \to \alpha + \pi$, λ, β then each value of h corresponds to four triples

$$\alpha, \ \lambda, \ \beta$$
$$\alpha + \pi, \ \lambda, \ \beta$$
$$\alpha, \ \lambda, \ \beta + \pi$$
$$\alpha + \pi, \ \lambda, \ \beta + \pi$$

where $-\frac{\pi}{2} \le \alpha < \frac{\pi}{2}$, $-\frac{\pi}{2} \le \beta < \frac{\pi}{2}$. Now we can write down the explicit expressions for (α, λ, β) in terms of elements of the matrix g (Eq.A.1):

$$\lambda^4 - (a^2 + b^2 + c^2 + d^2)\lambda^2 + 1 = 0 \tag{A.2}$$

$$\tan\alpha = \frac{b + \lambda^2 c}{d - \lambda^2 a} \tag{A.3}$$

$$\tan\beta = -\frac{a - \lambda^2 d}{b + \lambda^2 c} \tag{A.4}$$

Eq.(A.2) defines uniquely λ (for $\lambda > 1$).

Using the decomposition Eq.(A.1), we can represent the matrices g^N and g_{N+1} as follows

$$g^N = R_{\varphi(N)} D_{\lambda(N)} R_{\psi(N)}$$
$$g_{N+1} = R_{\varphi_{N+1}} D_{\lambda_{N+1}} R_{\psi_{N+1}}$$

Then the product $g_{N+1} \cdot g^N$ reads

$$g^{N+1} = g_{N+1} \cdot g^N = R_{\varphi_{N+1}} \left[D_{\lambda_{N+1}} R_{\psi_{N+1}+\varphi^{(N)}} D_{\lambda^{(N)}} \right] R_{\psi^{(N)}} \qquad \text{(A.5)}$$

Now we can put

$$\gamma^{(N)} = \psi_{N+1} + \varphi^{(N)}$$

and rewrite the product in square brackets in Eq.(A.5) as follows

$$D_{\lambda_{N+1}} R_{\gamma^{(N)}} D_{\lambda^{(N)}} = R_{\alpha^{(N)}} D_{\lambda^{(N+1)}} R_{\beta^{(N)}}$$

Then we have

$$\varphi^{(N+1)} = \varphi_{N+1} + \alpha^{(N)} \qquad (\text{mod } \pi) \qquad \text{(A.6)}$$
$$\psi^{(N+1)} = \psi^{(N)} + \beta^{(N)} \qquad (\text{mod } \pi) \qquad \text{(A.7)}$$

It is essential that $\lambda^{(N+1)}$ and $\varphi^{(N+1)}$ do not depend on $\psi^{(N)}$. Therefore their evolution with N may be considered independently. The exact recursion equation for $\lambda^{(N+1)}$ takes the form:

$$\left(\lambda^{(N+1)} \right)^4 - \left(\lambda^{(N+1)} \right)^2 \left[\left(\lambda_{N+1}^2 \left(\lambda^{(N)} \right)^2 + \frac{1}{\lambda_{N+1}^2 \left(\lambda^{(N)} \right)^2} \right) \cos^2 \gamma^{(N)} + \right.$$
$$\left. \left(\frac{\left(\lambda^{(N)} \right)^2}{\lambda_{N+1}^2} + \frac{\lambda_{N+1}^2}{\left(\lambda^{(N)} \right)^2} \right) \sin^2 \gamma^{(N)} \right] + 1 = 0$$

$$\text{(A.8)}$$

SIMPLIFYING ASSUMPTION. Suppose $\lambda^N \gg 1$ for $N > 1$. In this approximation Eq.(A.8) is replaced by

$$\left(\lambda^{(N+1)} \right)^4 - \left(\lambda^{(N+1)} \right)^2 \left[\lambda_{N+1}^2 \left(\lambda^{(N)} \right)^2 \cos^2 \gamma^{(N)} + \frac{\left(\lambda^{(N)} \right)^2}{\lambda_{N+1}^2} \sin^2 \gamma^{(N)} \right] = 0$$

$$\text{(A.9)}$$

which gives the solution

$$\lambda^{(N+1)} = \lambda^{(N)} \sqrt{\lambda_{N+1}^2 \cos^2 \gamma^{(N)} + \frac{1}{\lambda_{N+1}^2} \sin^2 \gamma^{(N)}} \qquad \text{(A.10)}$$

In the same approximation

$$\tan \alpha^{(N)} = \frac{1}{\lambda_{N+1}^2} \tan \gamma^{(N)} \qquad \text{(A.11)}$$

Now we are in position to formulate precisely our problem. Let us assume that the probability distribution $\mathcal{P}$ induces the distribution on $PSL(2,\mathbf{R})$ which will be denoted by the same letter. Thus, we can consider $\mathcal{P}$ as a probability distribution on the space of triples (φ, λ, ψ). Suppose we have N independent identically distributed triples $(\varphi_m, \lambda_m, \psi_m)$, each having the distribution $\mathcal{P}$. Consider the sequence of triples

$$\omega_m = (\Phi_m, \lambda_{m+1}, \varphi_{m+1}), \qquad 0 \le m \le N$$

where $\Phi_m \equiv \varphi^{(m)}$ is connected by Eqs.(A.6), (A.11) and has the initial condition $\varphi^{(0)} = 0$; λ_1, ψ_1 being arbitrary. Denote by $\mathcal{P}_N$ the induced probability distribution on the sequences of triples $\{\omega_N\}$.

Lemma A.1 *The probability distribution $\mathcal{P}_N$ is a Markov chain.*

Proof. Suppose we have a given pair (ω_{m-1}, ω_m). The equality (A.6) gives a possibility to write $\alpha^{(m-1)}$ as a function of φ_m, Φ_m, namely

$$\alpha^{(m-1)} = \Phi_m - \varphi_m \qquad (\text{mod } \pi)$$

Knowing $\alpha^{(m-1)}$ we can find $\gamma^{(m-1)}$ from Eq.(A.11) and ψ_m from the equality

$$\psi_m = \gamma^{(m-1)} - \varphi^{(m-1)} \qquad (\text{mod } \pi)$$

This shows that $\psi(\Phi_{m-1}, \lambda_m, \varphi_m, \Phi_m)$ is a single-valued inevitable function.

Write the density π corresponding to $\mathcal{P}$ in the form

$$\pi = \pi(\varphi, \lambda) \cdot \pi(\psi|\varphi, \lambda)$$

where $\pi(\psi|\varphi, \lambda)$ is the conditional density of ψ, provided that φ and λ are fixed. Then the density of the transitional probability distribution can be represented as

$$\pi(\omega_m|\omega_{m-1}) = \pi(\varphi_{m+1}, \lambda_{m+1}) \cdot \pi(\psi_m|\varphi_m, \lambda_m) \cdot \left| \frac{d\psi_m}{d\Phi_m} \right| \qquad (A.12)$$

It is clear that the conditional density $\pi(\omega_m|\omega_{m-1}, \omega_{m-2}, \ldots, \omega_1)$ depends evidently on ω_m and ω_{m-1} only. Thus it is proved that $\mathcal{P}_N$ is the Markov chain and its conditional transition density is found $\square$.

Eq.(A.10) shows that

$$\lambda^{(k+1)} = \lambda^{(k)} \exp\left\{ F(\omega_{k+1}, \omega_k) \right\} \qquad (A.13)$$

where

$$F\left(\omega_{k+1}, \omega_k\right) = \frac{1}{2} \ln \left(\lambda_{k+1}^2 \cos^2 \gamma^{(k)} + \lambda_{k+1}^{-2} \sin^2 \gamma^{(k)}\right) \qquad (A.14)$$

Rewriting Eqs.(A.13)-(A.14) in the form

$$\ln \lambda^{(m)} = \sum_{k=1}^{m} F\left(\omega_{k+1}, \omega_k\right) \qquad (A.15)$$

we can attribute to the function $\lambda^{(m)}$ the sense of the *Lyapunov exponent* of the product of the first m matrices of the product under consideration, $g^{(N)}$.

Take two numbers a_1 and a_2 and denote by $\mathcal{Q}_N$ conditional probability (induced by $\mathcal{P}_N$) of the fact that $a_1 \leq \ln \lambda^{(N)} \leq a_2$. It should be noted that we can drop the simplifying assumption (see above), so all $\lambda^{(m)}$ for $1 \leq m \leq N$ are no longer bigger than 1.

Thus we have all necessary definitions to be able to formulate our main problem. Fix a number ℓ, $0 < \varsigma < 1$, put $N_1 = [\varsigma N]$ and ask the following question. *What is the limit probability distribution for the value* $\frac{1}{\sqrt{N_1}} \ln \lambda^{(N_1)}$ *as* $N \to \infty$ where the distribution of $\lambda^{(N_1)}$ is determined by $\mathcal{Q}_N$.

We use the Cramer's method in the theory of probabilities of large deviations for the sequences of independent random variables. So, let us rewrite the density corresponding to the distribution P_N in the following form

$$\pi_0\left(\omega_0, \omega_1, \ldots, \omega_N\right) = \pi_0\left(\omega_0\right) \prod_{m=1}^{N} \pi\left(\omega_m | \omega_{m-1}\right) \qquad (A.16)$$

where $\pi\left(\omega_m | \omega_{m-1}\right)$ is the transitional density found in Lemma A.1. Now for any b $(-\infty < b < \infty)$ we introduce the new "normalized" probability distribution $\mathcal{P}_N(b)$ with the density

$$\pi_b\left(\omega_0, \omega_1, \ldots, \omega_N\right) = \frac{e^{b \sum_{m=1}^{N} F(\omega_m | \omega_{m-1})}}{\Theta_N(b)} \pi_0\left(\omega_0\right) \prod_{m=1}^{N} \pi\left(\omega_m | \omega_{m-1}\right) \qquad (A.17)$$

Using the statistical mechanics analogy we could say that the function in the exponent is the sum of "Boltzmann weights" of random steps with the "inverse temperature" $b \equiv -\frac{1}{T}$; hence $\Theta_N(b)$ is nothing else than the partition function determined by the normalization condition

$$\sum_{m=1}^{N} \pi_b \left(\omega_0, \omega_1, \ldots, \omega_N \right) = 1$$

Thus Eq.(A.17) defines the non-homogeneous Markov chain.

The joint probability distribution corresponding to Q_N reads

$$\overline{\pi}_b \left(\omega_0, \omega_1, \ldots, \omega_N \right) = \frac{\pi_0 \left(\omega_0 \right) \prod_{m=1}^{N} \pi \left(\omega_m | \omega_{m-1} \right)}{\Sigma(N)} \qquad (A.18)$$

where

$$\Sigma(N) = \int\limits_{\substack{\{\omega_0, \ldots, \omega_N\} \\ a_1 \le \lambda^{(N)} \le a_2}} \pi_0 \left(\omega_0 \right) \prod_{m=1}^{N} \pi \left(\omega_m | \omega_{m-1} \right) \prod_{m=0}^{N} d\omega_m \qquad (A.19)$$

Eq.(A.17) can be rewritten as follows

$$\pi \left(\omega_0, \ldots, \omega_N \right) = \pi_b \left(\omega_0, \ldots, \omega_n \right) \Theta_N(b) e^{-b \sum_{m=1}^{N} F(\omega_m, \omega_{m+1})}$$

The quantity $\Sigma(N)$ can be estimated from above

$$\Sigma(N) \le \Theta_N(b) e^{-ba_1} \int\limits_{\substack{\{\omega_0, \ldots, \omega_N\} \\ a_1 \le \lambda^{(N)} \le a_2}} \pi_b \left(\omega_0, \ldots, \omega_N \right) \prod_{m=0}^{N} d\omega_m \qquad (A.20)$$

and from below

$$\Sigma(N) \ge \Theta_N(b) e^{-ba_2} \int\limits_{\substack{\{\omega_0, \ldots, \omega_N\} \\ a_1 \le \lambda^{(N)} \le a_2}} \pi_b \left(\omega_0, \ldots, \omega_N \right) \prod_{m=0}^{N} d\omega_m \qquad (A.21)$$

Denote $\omega' = (\Phi', \lambda', \varphi')$, $\omega'' = (\Phi'', \lambda'', \varphi'')$ and consider the positive kernel

$$K_b(\omega'' | \omega') = \pi(\omega'' | \omega') e^{bF(\omega'' | \omega')}$$

associated with the integral operator

$$(K_b f) (\omega'') = \int K(\omega'' | \omega') f(\omega') d\omega' \qquad (A.22)$$

In turn, the operator adjoint to (A.22) has the following form

$$(K_b^* f)(\omega') = \int K(\omega''|\omega')f(\omega'')d\omega'' \qquad (A.23)$$

Lemma A.2 *The operator K_b has a positive eigenfunction*

$$hob(\omega') \equiv h_b(\Phi', \lambda', \varphi') = p(\varphi', \lambda')g_b(\Phi', \lambda', \varphi')$$

where g_b satisfies the following conditions:

1. $g_b(\Phi', \lambda', \varphi') = 0$ *if $p(\varphi', \lambda') = 0$;*

2. *for some positive constants c_1 and c_2 we have $c_1 < g_b(\Phi', \lambda', \varphi') < c_2$ if $p(\varphi', \lambda') > 0$. The adjoint operator K_b^* has a positive eigenfunction $h_b^*(\omega'') \equiv h_b(\Phi'', \lambda'', \varphi'')$ such that $c_1 < g_b^*(\lambda'', \varphi'') < c_2$*

3. *The corresponding eigenvalues of K_b and K_b^* coincide. We denote this common eigenvalue by $\Lambda(b)$.*

Proof. Let us begin by analysing of the eigenfunction h_b. Take an arbitrary function $u(\omega')$ in the form $u(\omega') = p(\varphi', \lambda')v(\Phi', \lambda', \varphi')$ where v is defined as follows $v = 0$ if $p(\varphi', \lambda') = 0$ and $d_1 < v(\Phi', \lambda', \varphi') < d_2$ in other cases; d_1 and d_2 are the positive constants.

Now from the definition of the kernel K_b and from Eq.(A.12) we have

$$(K_b u)(\omega'') = p(\varphi'', \lambda'') \int v(\Phi', \lambda', \varphi')p(\psi|\varphi', \lambda') \left| \frac{d\psi'}{d\varphi'} \right| e^{bF(\omega''|\omega')}d\Phi'$$

Put $v_1(\Phi, \varphi'', \lambda'') = 0$ if $p(\varphi'', \lambda'') = 0$ and

$$v_1(\omega'') = v_1(\Phi'', \varphi'', \lambda'') = \int v(\Phi', \varphi', \lambda')p(\psi|\varphi', \lambda') \left| \frac{d\psi'}{d\varphi'} \right| e^{bF(\omega'', \omega')}d\Phi',$$

then v_1 satisfies Condition 1 of Lemma A.2. We may assume that

$$0 < \frac{p(\psi'|\varphi', \lambda')}{p(\psi''|\varphi', \lambda')} \le d$$

for some d. Otherwise we can pass to the operator $(K_b)^s$, where $s = s(d)$.

For two different values $\overline{\omega}'' = \left(\overline{\Phi}'', \overline{\varphi}'', \overline{\lambda}''\right)$ and $\overline{\overline{\omega}}'' = \left(\overline{\overline{\Phi}}'', \overline{\overline{\varphi}}'', \overline{\overline{\lambda}}''\right)$ we have

$$
\frac{v_1\left(\overline{\omega}'\right)}{v_1\left(\overline{\overline{\omega}}''\right)} = \frac{\displaystyle\int v(\Phi', \varphi', \lambda') p(\overline{\psi}'|\varphi', \lambda') \left|\frac{d\overline{\psi}'}{d\varphi'}\right| e^{bF(\omega'', \omega')} d\Phi'}{\displaystyle\int v(\Phi', \varphi', \lambda') p(\overline{\overline{\psi}}|\varphi', \lambda') \left|\frac{d\overline{\overline{\psi}}}{d\varphi'}\right| e^{bF(\omega'', \omega')} d\Phi'}
$$

$$
\leq d \frac{\max \left|\dfrac{d\overline{\psi}'}{d\varphi'}\right|}{\min \left|\dfrac{d\overline{\overline{\psi}}}{d\varphi'}\right|} \exp\left\{b\left(\max F(\omega'', \omega') - \min F(\omega'', \omega')\right)\right\}
$$

Here $\overline{\psi}'$ and $\overline{\overline{\psi}}''$ are the values of ψ' corresponding to $\overline{\omega}''$ and $\overline{\overline{\omega}}''$ for the same ω'. If we put now $L_b v = v_1$, we immediately come to the conclusion that the operator L_b has a positive kernel on a compact set and therefore has a positive eigenfunction with the positive eigenvalue.

The same arguments work for the operator K_b^*. The fact that the operators K_b and K_b^* have the common set of eigenvalues can be shown by simple direct calculations $\square$.

Now we rewrite $\pi_b(\omega_0, \dots, \omega_N)$ (see Eq.(A.17)) as follows

$$
\pi_b(\omega_0, \dots, \omega_N) = \Lambda^N(b) \frac{\pi_0(\omega_0) g_b^*(\omega_0)}{\Theta_N(b) g^*(\omega_N)} \prod_{m=1}^N \frac{e^{bF(\omega_m, \omega_{m-1})} g_b^*(\omega_m)}{\Lambda(b) g_b^*(\omega_{m-1})}
$$

The function

$$
p_b(\omega'', \omega') = \frac{e^{bF(\omega'', \omega')} h_b^*(\omega'')}{\Lambda(b) h_b^*(\omega')}
$$

can be considered the kernel of a stochastic operator $\mathcal{P}_b$. The invariant measure for $\mathcal{P}_b$ has the density $\nu_b(\omega') = h_b(\omega') h_b^*(\omega')$. The functions g_b, g_b^* are normalized in such a way that $h_b(\omega')$ becomes the density of a probability measure. Thus, we have

$$
\Sigma(N) \leq \lambda^N(b) e^{-ba_1} \int_\Omega g_b^*(\omega_0) \pi_0(\omega_0) \prod_{m=1}^N p_b(\omega_m, \omega_{m-1}) \frac{1}{h^*(\omega_N)} \prod_{k=0}^N d\omega_k
$$

$$
\tag{A.24}
$$

and

$$\Sigma(N) \geq \lambda^N(b) e^{-ba_2} \int_\Omega g_b^*(\omega_0) \pi_0(\omega_0) \prod_{m=1}^{N} p_b(\omega_m, \omega_{m-1}) \frac{1}{h^*(\omega_N)} \prod_{k=0}^{N} d\omega_k$$

$$(A.25)$$

where Ω is the integration domain as in Eqs.(A.20), (A.21).

Lemma A.3 *There exists one and only one value b_0 for which*

$$\int F(\omega'', \omega') p_{b_0}(\omega'', \omega') \nu_{b_0}(\omega') d\omega d\omega'' = 0 \qquad (A.26)$$

Proof. This Lemma is well-known in statistical mechanics. It means that b_0 is found under the condition that the expectation of $F(\omega_1|\omega_0)$ with respect to the stationary Markov measure with transition density $p_{b_0}(\omega'', \omega')$ is equal to zero. It can be shown that

$$\frac{d}{db} \int F(\omega'', \omega') p_b(\omega'', \omega') \nu_b(\omega') d\omega' d\omega'' > 0 \qquad (A.27)$$

because we can rewrite the integral in Eq.(A.27) as follows

$$\int F(\omega'', \omega') p_b(\omega'', \omega') \nu_b(\omega') d\omega' d\omega'' =$$

$$\lim_{N\to\infty} \frac{1}{N} \frac{\partial}{\partial b} \ln \int \pi_0(\omega_0) \prod_{m=1}^{N} \pi(\omega_m|\omega_{m-1}) e^{b \sum_{m-1}^{N} F(\omega_m, \omega_{m-1})} \prod_{m=0}^{N} d\omega_m$$

$$(A.28)$$

and therefore

$$\frac{\partial}{\partial b} \int F(\omega'', \omega') p_b(\omega'', \omega') \nu_b(\omega') d\omega' d\omega'' =$$

$$\lim_{N\to\infty} \frac{1}{N} \mathrm{Var}_b \left(\sum_{m-1}^{N} F(\omega_m, \omega_{m-1}) \right)$$

where $\mathrm{Var}_b(\ldots)$ is the variance found with the help of distribution $\mathcal{P}_N(b)$. Thus Eq.(A.27) is proved. This yields that the expectation (Eq.(A.28)) is a monotone increasing function of b. It is easy to find such periodic sequences $\{\omega_m\}$ for which the sum $\sum_{m=1}^{t} F(\omega_m|\omega_{m-1})$ over a period t is strictly positive (or strictly negative). Correspondingly, the limit in Eq.(A.28) is positive as $b \to \infty$ and negative as $b \to -\infty$. Therefore there exists one and only one value of b_0 for which the integral Eq.(A.26) is zero $\square$.

Since

$$\ln \lambda^{(N)} = \sum_{m=1}^{N} F(\omega_m, \omega_{m-1})$$

we can use the local central limit theorem for Markov chains with a compact phase space which gives

$$\int\limits_{a_1 \leq \ln \lambda^{(N)} \leq a_2} g_{b_0}^*(\omega_0) \pi_0(\omega_0) \prod_{m=1}^{N} p_{b_0}(\omega_m, \omega_{m-1}) \frac{1}{h_{b_0}^*(\omega_N)} \prod_{k=0}^{N} d\omega_k \sim$$

$$\frac{1}{\sqrt{2\pi N \sigma}} \int g_{b_0}^*(\omega_0) \pi_0(\omega_0) h_{b_0}^*(\omega_N) d\omega_N$$

$$(A.29)$$

as $N \to \infty$. The constant $\sigma = \sigma(b) > 0$ enters in the asymptotics of the variance

$$\lim_{N \to \infty} \mathrm{Var}_b \left(\sum_{m-1}^{N} F(\omega_m, \omega_{m-1}) \right) \sim N\sigma$$

Take now two numbers u_1, u_2 $(u_1 < u_2)$ and consider the probability

$$q_N = Q \left\{ u_1 \leq \frac{1}{\sqrt{N_1}} \ln \lambda^{(N_1)} \leq u_2 \right\}$$

We have

$$q_N = \frac{1}{\Sigma(N)} \int\limits_{\overline{\Omega}} \pi_0(\omega_0) \prod_{m=1}^{N} \pi(\omega_m | \omega_{m-1}) \prod_{m=0}^{N} d\omega_m$$

$$\leq \frac{\Lambda_1^N(b) e^{-ba_1}}{\Sigma(N)} \int\limits_{\overline{\Omega}} \pi_0(\omega_0) g_{b_0}^*(\omega_0) \prod_{m=1}^{N} \pi_{b_0}(\omega_m | \omega_{m-1}) \frac{1}{g_{b_0}^*(\omega_N)} \prod_{k=0}^{N} d\omega_k$$

$$(A.30)$$

where $\overline{\Omega} = \left\{ \{\omega_0, \ldots, \omega_N\} : a_1 \leq \lambda^{(N)} \leq a_2; u_1 \leq \frac{1}{\sqrt{N_1}} \ln \lambda^{(N_1)} \leq u_2 \right\}$. The analogous inequality can be written estimating q_N from below. The probability density p_{b_0} follows from Eq.(A.30) and the local central theorem for Markov chains:

$$p_{b_0} \left\{ \ln \lambda^{(N_1)} = u \right\} \sim \frac{1}{\sqrt{2\pi N_1 \sigma}} \exp \left(-\frac{u^2}{2N_1 \sigma} \right)$$

For $u = O\left(\sqrt{N}\right)$ and $v = O\left(\sqrt{N}\right)$ the conditional probability density reads

$$p_{b_0}\left\{\ln \lambda^{(N)} = v | \ln \lambda^{(N_1)} = u\right\} \sim \frac{1}{\sqrt{2\pi(1-\varsigma)N\sigma}} \exp\left(-\frac{u^2}{2(1-\varsigma)N\sigma}\right)$$

This yields

$$\int_{\overline{\Omega}} \pi_0(\omega_0) h_{b_0}^*(\omega_0) \prod_{m=1}^{N} \pi_{b_0}(\omega_m | \omega_{m-1}) \frac{1}{h_{b_0}^*(\omega_N)} \prod_{k=0}^{N} d\omega_k \sim$$

$$\frac{1}{\sqrt{2\pi(1-\varsigma)\sigma}} \int_{u_1}^{u_2} e^{-\frac{u^2}{2(1-\varsigma)N\sigma}} du \frac{1}{\sqrt{2\pi N\sigma}} \int g_{b_0}^*(\omega_0)\pi_0(\omega_0)d\omega_0 \int g_{b_0}(\omega_N)d\omega_N$$

Combining Eqs.(A.24), (A.29), (A.30), we get

$$\frac{e^{b_0(a_2-a_1)}}{\sqrt{2\pi\sigma(1-\varsigma)}} \int_{u_1}^{u_2} e^{\frac{u^2}{2\sigma(1-\varsigma)}} du \le q_N \le \frac{e^{b_0(a_1-a_2)}}{\sqrt{2\pi\sigma(1-\varsigma)}} \int_{u_1}^{u_2} e^{\frac{u^2}{2\sigma(1-\varsigma)}} du \qquad (A.31)$$

Let us formulate now the main theorem:

Theorem A.1 (Sinai and Nechaev [2]) *The limit probability distribution of the variable $\frac{1}{\sqrt{N_1}} \ln \lambda^{(N_1)}$ with respect to the distribution $\mathcal{Q}_N$ is Gaussian with the variance $\sigma(1-\varsigma)$ and zero's mean.*

Proof. Take an interval $[a_1, a_2]$ and decompose it into small parts: $a_1 < a^{(1)} < a^{(2)} < \ldots < a^{(r)} = a_2$ such that $a^{(j)} - a^{(j-1)} \le e$ where $e > 0$ is a given small number. From Eq.(A.31) we have

$$\mathcal{P}\left\{u_1 \le \frac{1}{\sqrt{N_1}} \ln \lambda^{(N_1)} \le u_2 \,\Big|\, a_1 \le \lambda^{(N)} \le a_2\right\} =$$

$$\sum_{j=1}^{r} \frac{\mathcal{P}\{a^{(j-1)} \le \ln \lambda^{(N)} \le a^{(j)}\}}{\mathcal{P}\{a_1 \le \ln \lambda^{(N)} \le a_2\}}$$

$$\times \mathcal{P}\left\{u_1 \le \frac{1}{\sqrt{N_1}} \ln \lambda^{(N_1)} \le u_2 \,\Big|\, a^{(j-1)} \le \lambda^{(N)} \le a^{(j)}\right\} =$$

$$\frac{1}{\sqrt{2\pi\sigma(1-\varsigma)}} \int_{u_1}^{u_2} e^{-\frac{u^2}{2\sigma(1-\varsigma)}} du \sum_{j=1}^{r} \frac{\mathcal{P}\{a^{(j-1)} \le \ln \lambda^{(N)} \le a^{(j)}\}}{\mathcal{P}\{a_1 \le \ln \lambda^{(N)} \le a_2\}} \left[1 + \delta^{(j)}(N,\epsilon)\right]$$

where $\left|\delta^{(j)}(N,\epsilon)\right| \le 2\epsilon$ for sufficiently large N. This gives the desired result
□.

References

1. H Fürstenberg, Trans. Amer. Math. Soc., 198 (1963), 377
2. S.K. Nechaev, Ya.G. Sinai, Bol. Soc. Bras. Mat., 21 (1991), 121
3. V.N. Tutubalin, Prob. Theory and Appl., 10 (1965), 15; ibid 13 (1968), 65

Appendix B

POLYMER CHAIN IN RANDOM ARRAY OF
TOPOLOGICAL OBSTACLES

Although the thermodynamic properties of polymers in the regular lattices of topological obstacles have been the subject of multiple studies and researches, the statistics of loops in disordered arrays of immobile obstacles has not received all the attention it deserves. At the same time the absence of translational invariance in the last case might have some influence on statistics of polymer chains. In particular, this question has been recently analysed in papers [1] and [2] which we find reasonable to reproduce below in a slightly shortened form.

1.1. Phase Transition in Polymer Loops Induced by Topological Disorder

Lately much attention has been devoted to the investigation of equilibrium and dynamic properties of polymer chains with excluded volume in random environment in connection with valuable technical and biological applications. Among the most important problems are the problems of DNA gel-electrophoresis as well as of polymer filtration and adsorption in porous media [3,4]. However, as far as we know, all analytical works are concerned with investigation of disorder only in potential interactions [5,6,7,8,9], leaving topological constraints out of consideration. (To be more correct it should be stated that in numerical simulations [9] the volume interactions and entanglements were taken into account simultaneously). At the same time there is a set of problems where the topological disorder plays an exceptionally important role. First of all it concerns the swelling and the high-elasticity properties of networks prepared in concentrated solutions.

At present we distinguish in literature between two groups of works dealing with rubber and swelling properties of polymer networks with topological constraints.

In one group of works [10,11] the topological constraints have been modelled by the uniform lattice of rods nonintersectible for the chain. This problem has been solved analytically and the dependence of stress under strain for the uniaxial extension-compression obtained there is in compliance with corresponding experimental data [11]. But the principal shortcoming of this model is connected with the regularity of the lattice of obstacles. Below we show that the presence of disorder in distribution of obstacles radically changes the statistics of chains.

In another group of works [12,13] the main attention has been paid to the influence of topological state of network subchain and of preparation conditions on thermodynamic properties of gels. It is clear that fixed topological structure of polymer gel is the typical example of quenched disorder. Actually, the structure is formed during the network preparation and can not be changed without the network destruction. Because of strict topological constraints on chain conformations the full phase space of the gel is divided into separated domains—just like the multi–valley structure of the spin glass phase space. The application of replica approach allows the author of [12] to obtain reasonable results for rubber properties of networks. But the method elaborated there for identification of the topological state of the chains seems rather doubtful.

Here we consider the model of polymer loops with volume interactions randomly entangled with randomly distributed infinite parallel rods. Within the framework of field-theoretical approach we show that the last disorder changes essentially the statistics of the chain and under some conditions induces the collapse transition of the loop.

1.1.1. *Description of the System*

Consider the closed 3D–polymer chain with 2D–volume interactions topologically entangled with an array of immobile randomly distributed rods perpendicular to xy-plane. The density of the rods projected to xy-plane is assumed to be Gaussian (which we take instead of Poissonian one for simplicity). We take also the winding number (i.e. number of rods

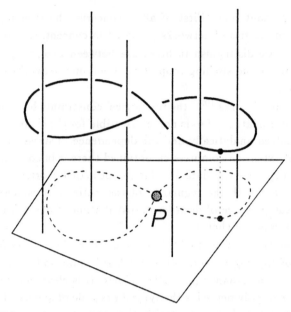

Fig. B.1. Polymer loop topologically entangled with random array of rods and its 2D projection where P is a point of contour selfintersection.

enclosed by the chain) to be quenched with Gaussian distribution having nonzero mean (see below).

The peculiarity of the volume interactions considered in our model consists in the fact that one chain segment excludes for the others not only one point in the space, but also the line perpendicular to the xy-plane. In other words all points of selfintersections in 2D projection are excluded. The typical conformation of closed chain in an array of rods in 3D space and its 2D projection with a selfintersection point P are shown in figs.B.1. Because of 2D character of volume interactions we avoid the points of selfintersections and deal with nonintersecting loops only. Since the disorder in positions of the rods is quenched we would like to calculate the free energy of the chain averaged over the distribution of the rods. Using the concept of the free energy self-averaging, the results obtained will give us the prediction of chain statistics in a given distribution of topological obstacles.

Some comments should be made concerning this model:

(i) We consider the chain attached in one point to the plane projection to avoid the situation of annealed disorder: actually, in absence of a fixed point and with no rods enclosed, the loop can visit all points of the volume sample. This indicates that the disorder in the distribution of obstacles changes effectively from quenched to annealed one;

(ii) We introduce such unusual volume interactions pursuing the forthcoming aims (see Section 1.2 of Appendix B). Namely, our final task concerns reconstruction of the true configuration of the closed chain with ordinary volume interactions embedded in 3D–space with points of selfintersections in the projection and topologically entangled with an array of rods. We will raise it from simple "blocks"—loops, considered here;

(iii) We discuss the possibility of collapse for different "preparation conditions" of the loop expecting that the interplay of disorders in winding number and in spatial distribution of rods can lead to the so-called "noise-induced" phase transition in polymer chain.

To the determine of the topological state of the closed chain with respect to the rods we can use the Gauss linking number. It is well known that this invariant is rather weak because it counts the number of obstacles entangled with the contour in *algebraic* sense only. For example, it can not distinguish from topological point of view the contour shown in fig.B.1 and the trivial unentangled one. But it seems to be obvious that in our model *just because of* 2D-*character of volume interactions and of the absence of contour selfintersections the Gauss invariant can be made complete.*

To analyze the equilibrium conformation of the loop in a random array of topological obstacles we use the mean-field approximation. That is, we choose the density of the loop's segments, ρ, as the order parameter and minimize the averaged free energy with respect to it. We find that the topological disorder induces the spontaneous symmetry breaking with $\rho \neq 0$ which we interprete as a collapsed phase of polymer loop.

1.1.2. *Field Representation of the Model*

Define the xy-coordinates of i's rod (obstacle) as $\mathbf{r}_i = (x_i, y_i)$. Then the Gauss invariant represented in terms of contour integral has the usual form [14]

$$G_i\{C\} = \frac{1}{2\pi} \oint_C ds \frac{\partial \mathbf{R}_\perp(s)}{\partial s} \nabla_\perp \left(\ln |\mathbf{R}_\perp(s) - \mathbf{r}_i| \right) \times \eta =$$

$$= \begin{cases} 1 & \text{point } \mathbf{r}_i \text{ is inside the contour } C \\ 0 & \text{otherwise} \end{cases} \tag{B.1}$$

where $\mathbf{R}(s)$ is the 3D radius-vector of the contour C, $\mathbf{R}_\perp(s)$ is its 2D-projection to the xy-plane, s is the variable along the contour and η is normal vector to xy-plane ($\eta = (0, 0, 1)$). The invariant $G_i(C)$ has the sense of the total angle (normalized by 2π) covered by the radius-vector $\mathbf{R}_\perp(s)$. This angle has the sign which depends on clock- or counter clockwise of path integration in Eq.(B.1).

Simple generalization of G_i of the form

$$I = \sum_{i=1}^{N} G_i\{C\} \tag{B.2}$$

counts the number of obstacles inside the contour C. It can be easily tested that the 2D-vector field $\mathbf{A}$ introduced as follows

$$\mathbf{A} = \sum_{i=1}^{N} \nabla_\perp \left(\ln |\mathbf{R}_\perp(s) - \mathbf{r}_i| \right) \times \eta \tag{B.3}$$

satisfies the conditions

$$\begin{aligned} \text{div}\,\mathbf{A} &= 0 \\ \text{rot}\,\mathbf{A} &= \eta[\varphi(\mathbf{R}_\perp) - \varphi_0] \end{aligned} \tag{B.4}$$

where $\varphi(\mathbf{R}_\perp) = \sum_i \delta(\mathbf{R}_\perp - \mathbf{r}_i)$ and φ_0 is the mean density of the obstacles in xy-plane. Then,

$$I = \oint \frac{\partial \mathbf{R}_\perp(s)}{\partial s} \mathbf{A} ds$$

We suppose $\varphi(\mathbf{R}_\perp)$ is randomly distributed with the Gaussian density $P_1\{\varphi(\mathbf{R}_\perp)\}$ which we write in the following form

$$\begin{aligned} P_1\{\varphi(\mathbf{R}_\perp)\} &\sim \exp\left\{ -\frac{1}{2\varphi_0} \int d\mathbf{R}_\perp \left[\varphi(\mathbf{R}_\perp) - \varphi_0\right]^2 \right\} \\ &= \exp\left\{ -\frac{1}{2\varphi_0} \int d\mathbf{R}_\perp (\text{rot}\,\mathbf{A})^2 \right\} \end{aligned} \tag{B.5}$$

where the dispersion φ_0 of the distribution $P_1\{\varphi(\mathbf{R}_\perp)\}$ coincides with the mean density.

The partition function of the loop containing inside fixed number of obstacles, c, reads

$$Z(c) = \int D\{\mathbf{R}\}\delta\left[\mathbf{R}(N) - \mathbf{R}(0)\right]\delta\left(c - \oint \frac{\partial \mathbf{R}_\perp(s)}{\partial s}\mathbf{A}ds\right)$$

$$\times \exp\left\{-\frac{1}{\ell^2}\oint\left(\frac{\partial \mathbf{R}(s)}{\partial s}\right)^2 ds - \frac{a^2}{2}\oint ds \oint ds'\delta\left(\mathbf{R}_\perp(s) - \mathbf{R}_\perp(s')\right)\right\}$$

(B.6)

where a^2 is the 2D-excluded volume and ℓ is the length of the chain segment.

In consideration the distribution in the number of obstacles inside the contour (below referred to as *topological charge*) we introduce the probability density $P_2(c)$ which is also assumed to be Gaussian (it is consistent with Eq.(B.1))

$$P_2(c) = \frac{1}{\sqrt{2\pi\Delta_c}}\exp\left\{-\frac{(c - c_0)^2}{2\Delta_c}\right\}$$

(B.7)

with mean c_0 and dispersion Δ_c.

We would like to calculate the free energy $F = -\ln Z$ of the loop averaged over the distributions Eq.(B.5) and Eq.(B.7). Although this problem is rather complex from the computational point of view, we can essentially simplify it by working not with topological charge, c, but with its chemical potential, g, having sense of the flux through the contour. Let us introduce the grand canonical ensemble, $Z(g)$, and the distribution, $P_2(g)$, defined as follows (see Eqs.(B.6) and (B.7)) correspondingly):

$$Z(g) = \int_{-\infty}^{\infty} e^{-icg}Z(c)dc$$

(B.8)

and

$$P_2(g) = \int_{-\infty}^{\infty} e^{-icg}P_2(c)dc = \left(\frac{\Delta_c}{2\pi}\right)^{1/2}\exp\left\{-\frac{\Delta_c}{2}\left(g + \frac{ic_0}{\Delta_c}\right)^2\right\}$$

(B.9)

For the mean free energy of the loop we have

$$<F> = -<\ln Z> = -\int P_1\{\varphi\}P_2(g)\ln Z(g)dg$$

(B.10)

Within the framework of the replica approach we exploit the identity

$$\langle F \rangle = -\frac{\partial}{\partial n} \langle Z^n(g) \rangle \Big|_{n=0} \tag{B.11}$$

where all calculations are performed for an integer value of n and the continuation of the value $n = 0$ is taken only in the final expression.

The replica partition function $Z^n(g)$ in the field representation has the form

$$Z^n(g) = \int \prod_{\alpha=1}^{n} D\{\psi_\alpha\} D\{\psi_\alpha^*\} \exp\left\{-\int d\mathbf{R} H^{(n)}(\Psi, \Psi^*, \mathbf{A}, g)\right\} \tag{B.12}$$

$$H^{(n)}(\Psi, \Psi^*, \mathbf{A}, g) = \sum_{\alpha=1}^{n} \psi_\alpha \left[\frac{\ell^2}{4}(\nabla_\perp - ig\mathbf{A})^2 + \frac{\ell^2}{2}\nabla_\parallel^2 + \tau\right]\psi_\alpha^*$$
$$+ \frac{La^2}{4}\left(\sum_{\alpha=1}^{n}\psi_\alpha\psi_\alpha^*\right)^2 \tag{B.13}$$

where $\Psi = \{\psi_1, \ldots, \psi_n\}$, $\Psi^* = \{\psi_1^*, \ldots, \psi_n^*\}$, τ is the chemical potential conjugate to the length of the loop, ℓ is the length of the chain segment and L is the mean size of the coil in z-direction.

Using the standard representation (see, for example, [15]), we can rewrite Eqs.(B.12)-(B.13) as follows

$$Z^n(g) = \frac{1}{\sqrt{\pi}} \int D\{\chi\} \exp\left\{-\frac{1}{2a^2}\int d\mathbf{R}\chi^2(\mathbf{R})\right\} \exp(-n\Phi) \tag{B.14}$$

where

$$\Phi = \ln \det\left[\frac{\ell^2}{4}(\nabla_\perp - ig\mathbf{A})^2 + \frac{l^2}{2}\nabla_\parallel^2 + \tau + i\chi\right] + \ln \pi \tag{B.15}$$

Eq.(B.11) becomes rather obvious in the graphical representation. After the integration over χ-variable it takes the form in fig.B.2 where each loop carries a factor n and due to the limit $n \to 0$ only the irreducible diagrams of type A remain. Each solid line in this expression corresponds to a polymer propagator and each dotted line represents the volume interaction.

In summary, it can be said that the continuation $n \to 0$ in the partition function $Z^n(g)$ (Eq.(B.14)) being applied to Eq.(B.11) makes it possible:

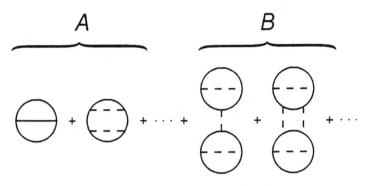

Fig. B.2. Illustration to Eq.(B.15)

- to average the logarithm of the partition function (i.e. the free energy) over the disorder;

- to extract the nonselfintersecting loops exclusively.

It should be emphasized, that in the case of ring chains the averaging of the free energy over the disorder as well as correct describing the volume interactions is governed simultaneously by one index n. From that point of view each field component should be considered as one replica.

Averaging of Eq.(B.12) over the spatial disorder (Eq.(B.5)) we obtain

$$\langle Z^n(g)\rangle = \int \prod_{\alpha=1}^{n} D\{\Psi, \Psi^*, \mathbf{A}\}\delta[\mathrm{div}\mathbf{A}]\exp\left\{-\int d\mathbf{R}\, H_{\mathrm{eff}}^{(n)}\left(\Psi, \Psi^*, \mathbf{A}, g\right)\right\} \tag{B.16}$$

where

$$H_{\mathrm{eff}}^{(n)}(\Psi, \Psi^*, \mathbf{A}, g) = \sum_{\alpha=1}^{n}\psi_\alpha\left[\frac{\ell^2}{4}\left(\nabla_\perp - ig\vec{A}\right)^2 + \frac{l^2}{2}\nabla_\parallel^2 + \tau\right]\psi_\alpha^*$$

$$+ \frac{1}{2\varphi_0}(\mathrm{rot}\mathbf{A})^2 + \frac{La^2}{4}\left(\sum_{\alpha=1}^{n}\psi_\alpha\psi_\alpha^*\right)^2 \tag{B.17}$$

The Hamiltonian Eq.(B.17) is nothing else than the Euclidean version of scalar electrodynamics. The correlator of $\mathbf{A}$-fields in the momentum space has the form

$$\langle \mathbf{A}_i(\kappa)\mathbf{A}_j(-\kappa)\rangle_0 = \frac{\varphi_0}{\kappa^2}\left(\delta_{ij} - \frac{\kappa_i\kappa_j}{\kappa^2}\right) \tag{B.18}$$

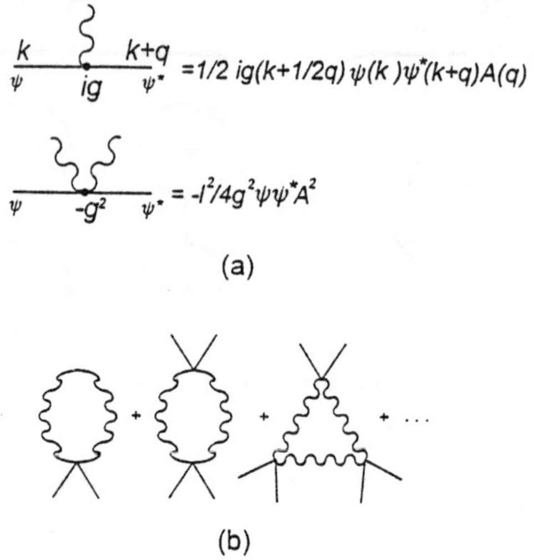

$$\frac{k \quad\quad k+q}{\psi \quad ig \quad \psi^*} = 1/2\ ig(k+1/2q)\ \psi(k)\ \psi^*(k+q)A(q)$$

$$\frac{}{\psi \quad -g^2 \quad \psi^*} = -l^2/4g^2\psi\psi^*A^2$$

(a)

(b)

Fig. B.3. a) Two types of vertex operators for the effective Hamiltonian Eq.(B.17); b) Graphic representation of the one-loop series contributing to the effective potential

1.1.3. *Effective Potential Method*

We analyze Eqs.(B.16)-(B.17) using the effective potential treatment which was used for the first time by S. Coleman and E. Weinberg [16] for similar Hamiltonian in quantum scalar electrodynamics. Considering the perturbative series in powers of magnetic flux, g, we can distinguish two kinds of vertices shown in fig.B.3a. The effective potential appears from the summation of one-loop diagrams with external ψ-lines taken at zero's momentum. Thus we have two possible series reproduced in fig.B.3b.

Following the arguments of Ref.[16] it can be shown that the series (a) vanishes because of the transversality of $\mathbf{A}$-correlator (Eq.(B.18)) and the remaining series (b) corresponds to the Gaussian integration in Eqs.(B.16)-(B.17).

Finally, in the one-loop approximation the effective Hamiltonian $H_{\text{eff}}^{(n)}$ can be rewritten as

$$H_{\text{eff}}^{(n)} = H_{\text{one-loop}}^{(n)} + \tau\Psi\Psi^* + \frac{La^2}{4}(\Psi\Psi^*)^2 + \text{const} \qquad (B.19)$$

where

$$H^{(n)}_{\text{one}-\text{loop}} = \frac{1}{2} \int \frac{d^2\kappa}{(2\pi)^2} \ln\left(\kappa^2 + Q\Psi\Psi^*\right); \qquad Q = \frac{\ell^2}{4}\varphi_0 g^2 \qquad (\text{B.20})$$

and

$$\Psi(\mathbf{R})\Psi^*(\mathbf{R}) = \sum_{\alpha=1}^{n} \psi_\alpha(\mathbf{R})\psi_\alpha^*(\mathbf{R})$$

Because the value of $H^{(n)}_{\text{one}-\text{loop}}$ is ultraviolet-divergent we cut off the integral in Eq.(B.20) at some value $\kappa = \Lambda$. Thus we obtain for $H^{(n)}_{\text{one}-\text{loop}}$ the expression:

$$H^{(n)}_{\text{one}-\text{loop}} = \frac{\Lambda^2}{8\pi}\left\{\ln\left[\Lambda^2 + Q\Psi\Psi^*\right] - 1\right\} - \frac{Q}{8\pi}\Psi\Psi^* \ln\frac{Q\Psi\Psi^*}{\Lambda^2 + Q\Psi\Psi^*} \qquad (\text{B.21})$$

Expanding Eq.(B.21) in power series in $Q\Psi\Psi^*/\Lambda^2$ and omitting the terms not depending on $Q\Psi\Psi^*$, we get

$$H^{(n)}_{\text{one}-\text{loop}} \simeq \frac{Q}{8\pi}\Psi\Psi^*\left(1 - \ln\left[\frac{Q}{\Lambda^2}\Psi\Psi^*\right]\right) \qquad (\text{B.22})$$

In compliance with the work [16] the replicated effective potential for our system can be written as follows

$$H^{(n)}_{\text{eff}} = \frac{Q}{8\pi}\Psi\Psi^*\left(1 - \ln\left[\frac{Q}{\Lambda^2}\Psi\Psi^*\right]\right) + \tau\Psi\Psi^* + \frac{La^2}{4}\left(\Psi\Psi^*\right)^2 + B\Psi\Psi^* \qquad (\text{B.23})$$

where we add the "mass" renormalization counterterm (the last term in Eq.(B.23)). It should be noted that because of the 2D character of the problem the coupling constant La^2 remains non-renormalized.

It should be emphasized that the $O(n)$ symmetry of the volume interaction term in Eq.(B.23) is illusory. Diagram series (fig.B.3a)) clearly show that only one-loop diagrams survive at $n \to 0$. The terms of the order of n^2 and higher are responsible for the interreplica interactions and under the condition $n \to 0$ we can neglect them with respect to the linear ones. So, the single-replica terms remain exclusively and the actual symmetry of the volume interaction term changes from $O(n)$ to the hypercubic one.

We define the renormalization of the mass, B, for the replica symmetric solution by the condition

$$\left.\frac{\partial^2}{\partial\psi\partial\psi^*}H^{(n)}_{\text{eff}}\right|_{\psi=\psi^*=M} = \tau n \qquad (\text{B.24})$$

where the subtraction point M is completely arbitrary. For the value of the counterterm B we have

$$B = \frac{Q}{8\pi} \ln \left(\frac{Q}{\Lambda^2} n M^2 \right) + \frac{Q}{8\pi} - L a^2 M^2 \qquad (B.25)$$

Substituting Eq.(B.23) for Eq.(B.25) we get

$$H_{\text{eff}}^{(n)} = -n \frac{Q}{8\pi} \psi \psi^* \ln \frac{\psi \psi^*}{M^2} + n \left(\tau + \frac{Q}{4\pi} - L a^2 M^2 \right) \psi \psi^* + n \frac{L a^2}{4} (\psi \psi^*)^2 \qquad (B.26)$$

The unusual logarithmic dependence of the mass on replica index disappears after renormalization.

The application of Eq.(B.11) to the partition function $\langle Z^n \{\psi \psi^*, g\} \rangle = \exp \left\{ -n \int d^3 r f_{\text{eff}} \{\psi \psi^*, g\} \right\}$ allows us to write the effective potential averaged over the spatial disorder in the distribution of the obstacles:

$$f_{\text{eff}} \left(|\psi|^2, g \right) = \tau^* |\psi|^2 - \frac{\ell^2}{32\pi} \varphi_0 g^2 |\psi|^2 \ln |\psi|^2 + \frac{L a^2}{4} |\psi|^4 \qquad (B.27)$$

The dependence on the cut-off parameter, M^2, is absorbed by the effective chemical potential τ^*. The value of τ^* controls the average number of chain segments, N, where $N = -\frac{\partial}{\partial \tau^*} \ln G(\tau^*, q = 0)$. The segment-to-segment correlation function in the loop, $G(\tau^*, q = 0)$, can be defined in the usual way:

$$G^{-1}(\tau^*, q = 0) = \frac{\partial^2}{\partial \psi \partial \psi^*} f_{\text{eff}} = \tau^*$$

Thus, in our approximation we have:

$$N = \frac{1}{\tau^*} \qquad (B.28)$$

Averaging Eq.(B.27) over the distribution, $P_2(g)$ we get

$$f_{\text{eff}} \{ |\psi|^2, \varphi, c_0, \Delta_c \} = \tau^* |\psi|^2 - \frac{\ell^2}{32\pi} \varphi_0 \frac{1}{\Delta_c} \left(1 - \frac{c_0^2}{\Delta_c} \right) |\psi|^2 \ln |\psi|^2 + \frac{L a^2}{4} |\psi|^4 \qquad (B.29)$$

Minimising of Eq.(B.29) with respect to the density $\rho = |\psi|^2$ we obtain

$$L a^2 \rho_0 = \frac{\ell^2}{16\pi} \varphi_0 \frac{1}{\Delta_c} \left(1 - \frac{c_0^2}{\Delta_c} \right) \left(1 + \ln(L^3 \rho_0) \right) - 2\tau^* \qquad (B.30)$$

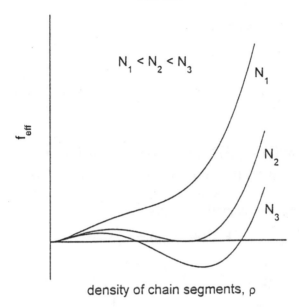

Fig. B.4. Dependence of the effective potential on the order parameter (i.e. density of chain segments) for different chain lengths.

This equation has one stable and one unstable solutions i.e. the loop undergoes the collapse phase transition of the first order—see fig.B.4. In particular, the binodal curve is determined by the equation

$$\frac{1}{N_c} = \frac{\ell^2}{32\pi}\varphi_0\frac{1}{\Delta_c}\left(1 - \frac{c_0^2}{\Delta_c}\right)\ln\left[\frac{\ell^2}{16\pi}\varphi_0\frac{1}{\Delta_c}\left(1 - \frac{c_0^2}{\Delta_c}\right)\left(\frac{L}{a}\right)^2\right] \quad (B.31)$$

where we have eliminated τ^* by using Eq.(B.28).

1.1.4. *The Role of Fluctuations*

Let us estimate the validity of the mean-field approximation considered in the previous section. We can easily expand the effective potential $f_{\text{eff}}\{\psi, \psi^*\}$ taking into account the fluctuations of the scalar fields ψ and

ψ^* near the stable mean-field solution:

$$f_{\text{eff}}\{\psi, \psi^*\} = f_{\text{eff}}\{\psi_0, \psi_0^*\}$$

$$\times \left[\frac{1}{2}\frac{\delta^2}{\delta\psi\delta\psi^*} f_{\text{eff}}\{\psi, \psi^*\}\Big|_{\psi=\psi_0 \ \psi^*=\psi_0^*} + k^2\lambda^2(\psi_0, \psi_0^*)\right]\Delta\psi\Delta\psi^*$$

(B.32)

where

$$\Delta\psi(k) = \psi(k) - \psi_0, \qquad \Delta\psi^*(k) = \psi^*(k) - \psi_0^*$$

In expansion Eq.(B.32) the last term in square brackets is described by the diagram series of the type (a) in figB.3b.

Performing the summation, we get

$$k^2\lambda^2(\psi_0, \psi_0^*, g) =$$

$$\left(\frac{\ell^2}{4}\right)^2 \bar{g}^2 \int \frac{d^2\kappa}{(2\pi)^2} \left(k_i + \frac{1}{2}\kappa_i\right)\left(k_j + \frac{1}{2}\kappa_j\right)\left(\delta_{ij} - \frac{\kappa_i\kappa_j}{\kappa^2}\right)$$

(B.33)

$$\times \frac{\varphi_0}{\kappa^2 + \frac{1}{4}L\ell^2\bar{g}^2|\psi_0|^2\varphi_0} \frac{1}{\frac{1}{4}\ell^2(k_\perp - \kappa)^2 + \frac{1}{2}\ell q^2 + \tau^*}$$

where $\bar{g}^2 = \frac{1}{\Delta_c}\left(1 - \frac{c_0^2}{\Delta_c}\right)$, $k_\perp$ and q are the components of the wave vector in xy-plane and along z-axis correspondingly.

It can be shown that near the binodal curve

$$\tau(|\psi_0|^2) \equiv \frac{\delta^2}{\delta\psi\delta\psi^*} f_{\text{eff}}\Big|_{\substack{\psi = \psi_0 \\ \psi^* = \psi_0^*}} = \tau_c^* - \tau^*$$

where τ^* is the value of the chemical potential on the binodal curve.

The mean square value of fluctuations of the fields ψ and ψ^* is

$$\langle|\Delta\psi|^2\rangle = \frac{1}{(2\pi)^2}\int_{|k|<l^{-1}} \frac{d^2k}{\tau(|\psi_0|^2)/2 + \lambda^2 k^2}$$

Performing the integration in Eq.(B.33) it can be seen that near the binodal curve the fluctuations weakness is determined by the inequality:

$$\frac{\langle|\Delta\psi|^2\rangle}{|\psi_0|^2} = \frac{a^2}{4\pi X\lambda_c^2(X)}\ln\frac{2\lambda_c^2(X)}{(\tau^* - \tau)l^2} \ll 1$$

where

$$X = \frac{\varphi_0 l^2 \bar{g}^2}{16\pi}$$

and the value λ^2 on the binodal curve is defined by the relation

$$\lambda_c^2(X) = \frac{l^2}{8\pi X(l/a)^2 - \ln(X(L/a)^2)} \ln \frac{2\pi X(l/a)^2}{\ln(X(L/a)^2)}$$

It should be emphasized that $X = (a/L)^2$ and $\lambda_c^2 \to \infty$ on the curve in fig.B.5. It means that the result obtained within the framework of the mean-field approach becomes exact on this line. Thus the mean-field approximation is valid in vicinity of the boundary of shaded area in fig.B.5.

1.1.5. *Comments and Conclusions*

We understand the collapse of the chain in the following sense. Take an ensemble of polymer loops of length N_0 with the mean number of obstacles enclosed c_0 and dispersion Δ_c. We call the set of values $(N_0, c_0, \Delta_c, \varphi_0)$ "the preparation conditions". For example, we could assume that the value c_0 appears in course of random closure of the chain of length N_0; $c_0 = c_0(N_0)$.

1. Let us grow the chain length *keeping the preparation conditions fixed*. It can be seen that when the loop reaches some critical value N_c, it becomes unstable and collapses according to Eq.(B.29). Actually, the spatial structure of long enough chains $(N \gg N_0)$ consists of entangled part of length N_1, bounded in the finite domain with dimension $D_0 = D_0(c_0, \varphi_0, \Delta_c)$ as well as of attached to it unentangled "octopus pulps-like" parts of common length $N - N_1$. The simple mean-field conjecture based on the exact calculation of the entropy of the loop unentangled in the array of obstacles (see, [17]) shows that loops of lengths $N > N_c$ form the collapsed state instead of increasing the dimensions of "octopus pulps-like" parts.

2. Let us discuss the role of dispersion Δ_c. We start with the case $c \neq 0$. It is clear that the dispersion in the number of obstacles inside the contour relieves the collapse transition, because some contours appear to be unentangled even in preparation conditions. From this point of view the equation on the binodal curve (Eq.(B.30)) can be interpreted as follows: for a fixed chain length, $N_c \gg N_0$, and spatial distribution of obstacles, φ_0,

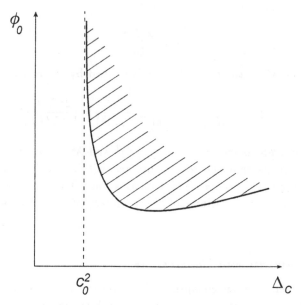

Fig. B.5. The phase diagram of the collapse transition. The shaded area corresponds to the collapsed state of the loop.

Eq.(B.30) gives the condition on the dispersion Δ_c necessary for producing phase transition (for the fixed mean value c_0).

Turn now to the case $c_0 = 0$. In this case all topological constraints are outside the chain. If we do not fix some point of the chain in the projection attached to the surface, the chain can move through the whole volume sample and the disorder should be regarded effectively as annealed. The fact that one point of the chain is kept fixed is not reflected in the mean density but it is so in correlation functions. We believe that in this case the microstructure of the chain resembles parts of condensed drops with tails between them (compare to [7]). Eq.(B.30) gives a reasonable answer to the case $c_0 = 0$: all contours are unentangled and the presence of dispersion (i.e., presence of part of entangled contours) only depresses the collapse transition.

The region of the values φ_0 and Δ_c, where the transition takes place is shown in fig.B.5 and is defined by the inequality

$$\frac{l^2}{16\pi}\varphi_0\frac{1}{\Delta_c}\left(1 - \frac{c_0^2}{\Delta_c}\right)\left(\frac{L}{a}\right)^2 > 1 \qquad (B.34)$$

It should be noted that the above mentioned mechanism of phase transition is a particular case of the so-called *noise-induced transitions* [17].

At least, for each nontrivial stable solution, ρ_0, of Eq.(B.30) the field **A** acquires a finite screening length, ξ_A:

$$\xi_A^{-1} \propto \varphi_0 l^2 \frac{1}{\Delta_c} \left(1 - \frac{c_0^2}{\Delta_c}\right) \rho_0 \qquad (B.35)$$

This is nothing else than the usual Higgs phenomenon [16]. In terms of our model it means that the loop has tried to be compactificated in the regions free of obstacles.

1.2. Classification and Statistics of Complex Loops

In this Section we extend the approach elaborated in Section 1.1 of Appendix B (Ref.[1]), hereafter referred to as I, to the description of thermodynamic properties of complex loops (i.e. loops with volume interactions and with points of chain selfintersections in xy-projection).

We would like to pay particular attention to investigation of the influence of "preparation condition" on thermodynamic properties of polymer loops. Under the "preparation conditions" we understand the initial topological state of the complex loop with respect to the randomly distributed array of parallel rods. The importance of this question is obvious. Elastic and swelling properties of polymer networks and gels depend strongly on the initial topological configuration of chains in the sample. The corresponding experimental data can be found in [20] whereas [21] is devoted to some qualitative physical explanations of this phenomenon.

Let us consider the following facts:

(i) An arbitrary complex loop in a quenched array of topological obstacles can be represented in a "cactus-like form" with "leaves" being the simple loops containing no points of selfintersections on the projection to the xy-plane. The points fastening different simple loops together function as the points of selfintersections.

(ii) The collapse transition of the complex loop with fixed preparation conditions (see below) can occur independently in different leaves when the chain length is increased, though the simultaneous collapse in all leaves is entropically unfavourable.

The methods used here have been proposed in I and represent a combination of the field theoretical effective potential treatment [16] with the

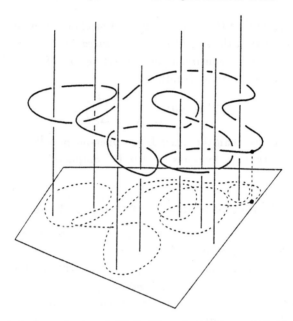

Fig. B.6. Example of complex loop (with its 2D projection) entangled with random array of topological obstacles.

replica approach [9]. The instability with respect to the collapse transition in different leaves can be described in a self-consistent way by the so-called asymmetric solution corresponding to the simplest (Gaussian) trial function with a variational parameter. This method corresponds to the well known Feynman variational principal [23].

1.2.1. *"Cactus-like" Representation of Complex Loop*

Consider an arbitrary polymer loop embedded in 3D space and topologically entangled with an array of immobile randomly distributed rods normal to xy-plane (see fig.B.6). We assume that the loop does not produce any entanglements and that the difference between the model under consideration and that shown in fig.B.1 consists in presence of many points of selfintersections in xy-projection in fig.B.6.

Let us neglect for a moment the topological constraints and pay attention to the *shadow graph*, obtained by projection of the closed chain onto the xy-plane. Assume that this graph is in general position i.e. con-

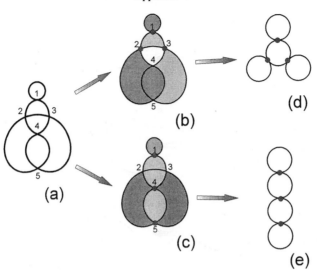

Fig. B.7. Shadow graph (a); its representation by two cactus–like graphs [(b) and (c)] and their topologies [(d) and (e)].

tains double points of path selfintersection only. The triple and higher order points of selfintersections can be removed by means of infinitesimal continuous deformation of the path. An arbitrary shadow graph can be represented in form of equivalent cactus-like graphs where each leaf has no points of selfintersections.

This correspondence is not unique as it can be seen from an example shown in fig.B.7: one and the same shadow graph can be represented by topologically different cactus-like graphs (see fig.B.7b,c). Thus it is necessary to distinguish between two types of points in cactus-like graphs:

(i) Points joining different cactus leaves. They are called *junctions* and are marked by bold circles (points 1,2,3 in fig.B.7b and points 1,4,5 in fig.B.7c);

(ii) Points of overlapping of different leaves. They are called *regular points* (points 4,5 in fig.B.7b and points 2,3 in fig.B.7c).

The difference in representation of the complex loop by shadow graphs corresponds to different ways of preparation of the initial shadow graph. The way of preparation is a succession of "elementary technological operations" which produce a complex loop with junctions and regular points

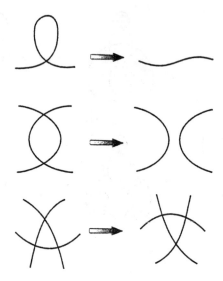

Fig. B.8. Reidemeister moves for shadow diagrams.

starting from a loop without selfintersections. To be more rigorous let us introduce the following definition.

Definition B.1 *We distinguish between the Reidemeister moves (see Chapter 1) by establishing the equivalence relations of shadow link diagrams. According to our nomenclature the crossing point in the move* I *is the junction whereas the crossing points in moves* II *and* III *are the regular points—see fig.B.8.*

Statement 5 *The regular way of preparation of an arbitrary shadow graph corresponding to any complex loop is as follows:*

(1) *Take a simple loop and create the necessary number of junctions by means of Reidemeister move of type* I.

(2) *Deform continuously the resulting cactus-like graph employing Reidemeister moves* II *and* III *keeping the number of junction points fixed.*

(3) *Sum over all topologically different cactus-like graphs resulting in the complex loop.*

Conjecture 6 *We suggest the probability distribution of formation of different cactus-like graphs to be uniform.*

In this case the fraction of shadow graphs in ensemble is increased proportionally to the number of its different representations by the cactus-like graphs. For example, the fraction of graphs like the one shown in fig.B.7a is increased twofold being a result of the two types of its cactus-like covering. So, this assumption gives us the possibility to determine the correspondence between ensembles of all shadow graphs and all cactus-like representations. Otherwise, for the nonuniform probability distribution it would be necessary to discriminate fractions of specific cactus-like graphs (e.g., fig.B.7d and fig.B.7e) which makes calculations much more complicated but does not change the main conclusion of the Section. It should be emphasized that the uniform probability distribution has nothing in common with the actual statistical weight of cactus-like graphs, resulting from the configurational partition function of a graph with a fixed number of junction points. *The real form of the probability distribution is the problem of preparation conditions of the complex loop.*

Finally, we emphasize that in principle different preparation conditions could correspond to the cactus-like graphs of the same topology (see, for instance fig.B.9). From the topological point of view the graphs in fig.B.9b and fig.B.9c are identical, both of them should contribute to the shadow graph (fig.B.9a). To distinguish between these equivalent graphs we have to define from very beginning the orientation (i.e. the direction of pathway) on the shadow graph. Hence the topological charge of a simple loop (leaf) can be both positive and negative (as it is in I).

1.2.2. *Effective Hamiltonian and Mean-field Free Energy of Complex Loops*

The Hamiltonian describing the complex loop can be written as follows (compare to Eq.(B.17))

$$H^{(n,m)}(\Psi, \Psi^*, \mathbf{A}, g) = \sum_{\alpha=1}^{n} \sum_{i=1}^{m} \psi_{\alpha i} \left(\frac{\ell^2}{4}(\nabla_\perp - ig\mathbf{A})^2 + \frac{\ell^2}{2}\nabla_\parallel^2 + \tau \right) \psi_{\alpha i}$$

$$+ \frac{La^2}{4} \sum_{\alpha=1}^{n} (\psi_{\alpha i}\psi_{\alpha i}^*)^2 - \frac{w}{4m} \sum_{\alpha=1}^{n} \sum_{i \neq j}^{m} \psi_{\alpha i}\psi_{\alpha i}^*\psi_{\alpha j}\psi_{\alpha j}^* + \frac{1}{2\varphi_0}(\nabla \times \mathbf{A})^2$$

$$(B.36)$$

To discriminate the complex loop structure, we have attributed an additional index i $(i = 1, \ldots, m)$ to the components of the fields Ψ and Ψ^*

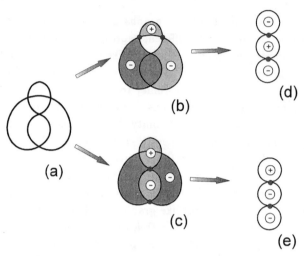

Fig. B.9. Twofold [(b) and (c)] cactus–like representation of shadow graph (a); [diagrams (d) and (e) are topologically equivalent to each others but have different distribution of topological charges.

enumerating different leaves. The interaction constant w stands for the fugacity of the number of junction points connecting different leaves; τ is the mass term, i.e. the chemical potential conjugated to the length of the whole connected graph, $\{\mathbf{A},\ g,\ \varphi_0,\ L,\ \ell,\ a^2\}$ are the same as in Eq.(B.17). The Greek indices are attributed now to the whole connected graph.

The first fourth-order term in Eq.(B.36) corresponds to the 2D excluded volume interactions inside a simple loop (as in I); whereas the second fourth-order term fastens together different simple loops.

The partition function corresponding to the Hamiltonian Eq.(B.36) is given by

$$Z^{(n,m)} = \exp\{-F^{(n,m)}\} = \int D\mathbf{A}\,D\psi_{\alpha i}\,D\psi^*_{\alpha i}\,\exp\Big\{-H\{\psi_{\alpha i};\psi^*_{\alpha i},\mathbf{A}\}\Big\} \tag{B.37}$$

The generating function of the connected graphs (the free energy) has the form

$$F^{(n,m)}\{\tau,w\} = n\sum_{k=1}^{m}\sum_{p=1}^{\infty} C_m^k\,\Im(k;p)\left(-\frac{w}{m}\right)^p \tag{B.38}$$

where $\Im(k;p)$ is a contribution of a connected graph constructed from k simple loops with p junction points and the combinatorial factor C_m^k gives the number of ways in which such contribution can be accomplished.

It is easy to see from Eq.(B.38) that in the limit $m \rightarrow \infty$ only the cactus-like graphs with $p = k - 1$ survive. As a result the free energy of the quenched cactus-like system is given by

$$F_{cactus} = \lim_{n \to 0} \lim_{m \to \infty} \frac{1}{m} \frac{\partial}{\partial n} Z^{(n,m)} = \sum_{k=1}^{\infty} \mathcal{P}(k; k-1) \frac{(-w)^{k-1}}{k!} \qquad (B.39)$$

where $(mn) \rightarrow 0$ and the factor $k!$ appears from the fact that the junction points are indistinguishable.

Now we can perform integration in Eq.(B.38) over the vector-potential **A** (as in I). The effective free energy in the mean-field approximation reads

$$F^{(n,m)}(\tau, w) = \frac{La^2}{4} \sum_{\alpha=1}^{n} (\psi_{\alpha i} \psi_{\alpha i}^*)^2 - \frac{w}{4m} \sum_{\alpha=1}^{n} \sum_{i \neq j}^{m} \psi_{\alpha i} \psi_{\alpha i}^* \psi_{\alpha j} \psi_{\alpha j}^*$$

$$- \frac{\varphi \ell^2 \bar{g}^2}{32\pi} \left(\sum_{\alpha=1}^{n} \sum_{i=1}^{m} \psi_{\alpha i} \psi_{\alpha i}^* \right) \left\{ 1 - \ln \left[\frac{\varphi^2 l \ell^2 \bar{g}^2}{4\Lambda^2} \left(\sum_{\alpha=1}^{n} \sum_{i=1}^{m} \psi_{\alpha i} \psi_{\alpha i}^* \right) \right] \right\}$$

$$+ \tau \sum_{\alpha=1}^{n} \sum_{i=1}^{m} \psi_{\alpha i} \psi_{\alpha i}^* + B \sum_{\alpha=1}^{n} \sum_{i=1}^{m} \psi_{\alpha i} \psi_{\alpha i}^*$$

$$(B.40)$$

where

$$\bar{g}^2 = \frac{1}{\Delta_c} \left(1 - \frac{c_0^2}{\Delta_c} \right) \qquad (B.41)$$

c_0 and Δ_c stand for the mean topological charge of simple loops and its dispersion. In Eq.(B.40) Λ is a cut-off parameter which has appeared due to the ultraviolet divergence of a resulting integral and the mass renormalization counterterm, B, is introduced to subtract this divergence. Thus one has

$$\frac{\partial^2}{\partial \psi_{\alpha i} \partial \psi_{\alpha i}} F^{(n,m)} \bigg|_{\psi_{\alpha i} = \mu_{\alpha i}; \ \psi_{\alpha i}^* = \mu_{\alpha i}^*} = \tau \delta_{\alpha\beta} \delta_{ij} \qquad (B.42)$$

with an arbitrary subtraction point $\psi_{\alpha i} = \mu_{\alpha i}$, $\psi_{\beta j}^* = \mu_{\beta j}^*$. Let we assume this point to be symmetrical with respect to all indices

$$\mu_{\alpha i} = \mu_{\alpha i}^* \equiv \mu \qquad (B.43)$$

Using Eqs.(B.42) and (B.43) we arrive at the following expression for the free energy

$$F^{(n,m)}(\tau,w) = \sum_{\alpha=1}^{n}\sum_{i\neq j}^{m} V_{ij}\psi_{\alpha i}\psi_{\alpha i}^{*}\psi_{\alpha j}\psi_{\alpha j}^{*}-$$

$$\frac{\varphi\ell^2\bar{g}^2}{32\pi}\left(\sum_{\alpha=1}^{n}\sum_{i=1}^{m}\psi_{\alpha i}\psi_{\alpha i}^{*}\right)\ln\left[\frac{1}{mn\mu^2}\left(\sum_{\alpha=1}^{n}\sum_{i=1}^{m}\psi_{\alpha i}\psi_{\alpha i}^{*}\right)\right]+$$

$$\tau\sum_{\alpha=1}^{n}\sum_{i=1}^{m}\psi_{\alpha i}\psi_{\alpha i}^{*}-2\mu^2\sum_{\alpha=1}^{n}\sum_{i\neq j}^{m}V_{ij}\psi_{\alpha i}\psi_{\alpha j}^{*}+$$

$$\frac{\varphi\ell^2\bar{g}^2}{32\pi}\left(\frac{1}{mn}\sum_{\alpha\neq\beta}^{n}\sum_{i\neq j}^{m}\psi_{\alpha i}\psi_{\alpha j}^{*}+\sum_{\alpha\neq\beta}^{n}\sum_{i}^{m}\psi_{\alpha i}\psi_{\alpha i}^{*}\right)$$

(B.44)

where the fourth-order vertex constant V_{ij} has the form

$$V_{ij} = \frac{L\,a^2}{4}\delta_{ij} - \frac{w}{4m}(1-\delta_{ij})$$

(B.45)

The validity of the mean-field approximation and the role of the intrareplica fluctuations have been investigated in I. The contribution of interreplica fluctuations as well as the possibility of replica symmetry breaking are much more involved problems which need additional investigation and are beyond the scope of the present analysis.

1.2.3. *Variational Principle*

The Eq.(B.44) for the effective mean-field free energy is symmetric with respect to the permutations of the ψ fields in the leaves and replica spaces. On the other hand the nontrivial structure of the fourth-order vertex (Eq.(B.45)) in the leaf space forces us to look for an asymmetric leave-space-solution in the condensed state. Such solution (symmetric in the replica space and asymmetric in the leaf space) can be written in the form

$$\psi_{\alpha i} = \overline{\psi} + \theta_i; \qquad \psi_{\alpha i}^{*} = \overline{\psi}^{*} + \theta_i^{*}$$

(B.46)

where $\overline{\psi}$ and $\overline{\psi}^{*}$ are the average values of the order parameters and θ fields determine the dispersion in the leaf space. Let us suggest for θ, θ^{*} the trial function in the Gaussian form:

$$P(\theta_i,\theta_i^{*}) \propto \exp\left\{-\frac{\theta_i\theta_i^{*}}{2\sigma^2}\right\}$$

(B.47)

where

$$\sigma^2 = \frac{1}{m} \sum_{i=1}^{\infty} \theta_i \theta_i^* \tag{B.48}$$

The distribution Eq.(B.47) can be regarded as test function with the variational parameter σ^2.

Now we are able to utilize variational principle. Substitute Eq.(B.44) for Eqs.(B.46) and (B.48) taking into account Eq.(B.47) and proceed with the limit $m \to \infty$. As a result we obtain the free energy in the asymmetric state

$$f_a(\rho_a, \sigma^2, \tau, w) = \lim_{n \to 0} \lim_{m \to \infty} \frac{1}{mn} F^{(n,m)} \{\rho_a, \sigma^2\} \tag{B.49}$$

where $\rho = \psi \psi^*$ stands for the density. The variational parameters ρ and σ^2 are the solution of pair of coupled equations

$$\frac{\partial}{\partial \rho_a} f_a(\rho_a, \sigma^2) = 0; \qquad \frac{\partial}{\partial \sigma_2} f_a(\rho_a, \sigma^2) = 0 \tag{B.50}$$

The asymmetric solution corresponds to the case $\sigma^2 > 0$. It becomes symmetric for $\sigma^2 = 0$:

$$f_s(\rho_s, \sigma^2 = 0) = \lim_{n \to 0} \lim_{m \to \infty} \frac{1}{mn} F^{(n,m)}(\rho_s, \sigma^2 = 0) \tag{B.51}$$

in this case the free energy is only a function of the density. Minimizing $f_s(\rho_s, \sigma^2 = 0)$ with respect to ρ_s we find the density of the symmetric state considered in I.

Performing the calculations described above we find

$$
\begin{aligned}
f_a(\rho, \sigma^2) &= \frac{1}{4}(La^2 - w)\rho^2 + \left(La^2 - \frac{w}{2}\right)\sigma^2\rho + \frac{1}{2}\left(La^2 - \frac{w}{2}\right)(\sigma^2)^2 \\
&\quad - \frac{\varphi \ell^2 \bar{g}^2}{32\pi}(\rho + \sigma^2)\ln\frac{\rho + \sigma^2}{\mu^2} + \tau(\rho + \sigma^2) - \mu^2(La^2 - w)\rho \\
&\quad + \mu^2\left(La^2 - \frac{w}{2}\right)\sigma^2 \frac{\varphi \ell^2 \bar{g}^2}{16\pi}\rho + \frac{\varphi \ell^2 \bar{g}^2}{32\pi}\sigma^2
\end{aligned}
\tag{B.52}
$$

A. SYMMETRIC SOLUTION. In this case $\sigma^2 = 0$ and the free energy takes the form

$$f_s(\rho_s, \tau, w) = \frac{1}{4}(La^2 - w_s)\rho_s^2 - \frac{\varphi \ell^2 \bar{g}^2}{32\pi}\rho_s \ln\frac{\rho_s}{\mu^2} + \tau_s^* \rho_s \tag{B.53}$$

where

$$\tau_s^* = \tau - \mu^2 (La^2 - w_s) + \frac{\varphi \ell^2 \bar{g}}{16\pi} \tag{B.54}$$

The minimization of Eq.(B.53) with respect to ρ_s yields

$$(La^2 - w_s)\rho_s = \frac{\varphi \ell^2 \bar{g}^2}{16\pi} \ln \left(\frac{\rho_s}{\mu^2} + 1 \right) - 2\tau_s^* \tag{B.55}$$

As discussed in I this equation has one stable and one unstable solution. Hence the first-order phase transition (loop condensation) occurs and the binodal curve is determined by the equation

$$\frac{1}{N_s^{\text{bin}}} = \frac{\varphi \ell^2 \bar{g}^2}{32\pi} \ln \left(\frac{\varphi \ell^2 \bar{g}^2}{16\pi} \frac{L}{(La^2 - w_s)\mu^2} \right) \tag{B.56}$$

where N_s^{bin} stands for the renormalized loop length at the binodal point. We define (as in I)

$$N = \frac{1}{\tau^*}$$

It is obvious that the symmetric solution corresponds to the case of the simple loop considered in I with the replacement $La^2 \rightarrow La^2 - w_s$. We have $w_s > 0$ what means that $N_s^{\text{bin}} < N_{\text{simple loop}}^{\text{bin}}$.

The free energy Eq.(B.53) at the binodal point has the form

$$f_s^{\text{bin}} = \frac{1}{4} \left(\frac{\varphi \ell^2 \bar{g}^2}{16\pi} \right)^2 \frac{1}{La^2 - w_s} \tag{B.57}$$

B. ASYMMETRIC SOLUTION. In this case the Eqs.(B.50) takes the form

$$\frac{1}{2}(La^2 - w_a)\rho_a + \left(La^2 - \frac{w_a}{2} \right) \sigma_a^2 - \frac{\varphi \ell^2 \bar{g}^2}{32\pi} \ln \frac{\rho_a + \sigma_a^2}{\mu^2}$$
$$+ \tau - \mu^2 (La^2 - w_a) + \frac{\varphi \ell^2 \bar{g}^2}{32\pi} = 0 \tag{B.58}$$

and

$$\left(La^2 - \frac{w_a}{2} \right) \rho_a + \left(La^2 - \frac{w_a}{2} \right) \sigma_a^2 - \frac{\varphi \ell^2 \bar{g}^2}{32\pi} \ln \frac{\rho_a + \sigma_a^2}{\mu^2}$$
$$+ \tau - \mu^2 \left(La^2 - \frac{w_a}{2} \right) = 0 \tag{B.59}$$

These equations can be transformed into:

$$(2La^2 - w_a)\sigma_a^2 = \frac{\varphi\ell^2\overline{g}^2}{16\pi}\left(\ln\frac{\rho_a + \sigma_a^2}{\mu^2} + 1\right) - 2\tau_a^* - (La^2 - w_a)\rho_a \quad (B.60)$$

where

$$La^2\rho_a = \frac{\varphi\ell^2\overline{g}^2}{16\pi} + \mu^2 w_a \quad (B.61)$$

and

$$\tau_a^* = \tau - \mu^2(La^2 - w_a) + \frac{\varphi\ell^2\overline{g}^2}{16\pi} \quad (B.62)$$

The analysis of these equations shows again that the nontrivial values of ρ and σ^2 appear as a result of the first order phase transition. For binodal curve we find

$$\frac{1}{N_a^{\text{bin}}} = \frac{\varphi\ell^2\overline{g}^2}{32\pi}\ln\left(\frac{\varphi\ell^2\overline{g}^2}{16\pi}\frac{L}{(2La^2 - w_a)\mu^2}\right) + \frac{1}{2}La^2\rho_a \quad (B.63)$$

The corresponding free energy (Eq.B.52)) at the binodal point has the form

$$f_a^{\text{bin}} = \frac{1}{4}\left(\frac{\varphi\ell^2\overline{g}^2}{16\pi}\right)^2\frac{1}{2La^2 - w_a} + \frac{1}{4}La^2\rho_a \quad (B.64)$$

Finally let us extract the density of the junction points, κ. In the symmetric case we have

$$\kappa = w_s\frac{\partial f_s^{\text{bin}}}{\partial w_s} = \frac{w_s}{4}\left(\frac{\varphi\ell^2\overline{g}^2}{16\pi}\right)^2\frac{1}{(La^2 - w_s)^2} \quad (B.65)$$

whereas in the asymmetric case we have

$$\kappa = w_a\frac{\partial f_a^{\text{bin}}}{\partial w_a} = \frac{w_a}{4}\left(\frac{\varphi\ell^2\overline{g}^2}{16\pi}\right)^2\frac{1}{(2La^2 - w_a)^2} \quad (B.66)$$

The expression of difference between N_a^{bin} and N_s^{bin} follows directly from Eqs.(B.56) and (B.63):

$$\frac{1}{N_a^{\text{bin}}} - \frac{1}{N_s^{\text{bin}}} = \frac{\varphi\ell^2\overline{g}^2}{16\pi}\left(\ln\frac{La^2 - w_s}{2La^2 - w_a} + 1\right) + \mu^2 w_a \quad (B.67)$$

If the densities of junctions in symmetric and asymmetric cases are equal, we can write

$$\frac{La^2 - w_s}{2La^2 - w_a} = \frac{w_s}{w_a} \quad (B.68)$$

where

$$w_s = \frac{1}{2}(2La^2 + Y) - \sqrt{La^2Y + \frac{1}{4}Y^2}$$
$$w_a = \frac{1}{2}(4La^2 + Y) - \sqrt{2La^2Y + \frac{1}{4}Y^2} \quad ; \quad Y = \frac{1}{4\kappa}\left(\frac{\varphi\ell^2\bar{g}^2}{16\pi}\right)^2 \quad \text{(B.69)}$$

Supposing the density of junction points to be high, we get

$$\frac{La^2\kappa}{\left(\varphi_0\ell^2\bar{g}^2\right)^2} \gg 1 \qquad \text{(B.70)}$$

In this region Eq.(B.67) takes the form

$$\frac{1}{N_a^{\text{bin}}} - \frac{1}{N_s^{\text{bin}}} \approx \frac{\varphi\ell^2\bar{g}^2}{16\pi}(1 - \ln 2) + \mu^2 w_a > 0 \qquad \text{(B.71)}$$

and we arrive at the following inequality

$$N_a^{\text{bin}} < N_s^{\text{bin}} < N_{\text{simple loop}}^{\text{bin}} \qquad \text{(B.72)}$$

1.2.4. *Comments and Conclusions*

Eq.(B.72) has very clear meaning. Let us take an ensemble of complex loops, where each chain has the length N_0, carries the topological charge with the mean value c_0 and dispersion Δ_c and has the density of junction points κ. The distribution of obstacles is characterized by its 2D-density φ_0. Call the set $\{N_0, c_0, \Delta_c, \kappa, \varphi_0\}$ the *preparation conditions* (compare to I, where $\kappa = 0$). If we increase the chain length keeping the preparation conditions fixed then the loop condensation occurs at $N = N_a^{\text{bin}}$ and the equilibrium thermodynamic state of the loop becomes asymmetric in leaf sizes, i.e. from geometrical point of view the complex loop resembles one long simple loop with many small "subloops" attached.

We emphasize that the inequality Eq.(B.70) is crucial for the Eq.(B.72). It is easy to see that in opposite limit in Eq.(B.70) (small density of junction points), Eqs.(B.60)-(B.62) lead to an unphysical solution $\sigma^2 < 0$. This means that the asymmetric state becomes favourable at high density of junctions or at highly entangle state. It follows from Eq.(B.69) that in this region $w_s < La^2$ and $w_a < 2La^2$, so the fourth-order vertex does not change its sign and the collapse is induced by the random distribution of obstacles only.

We have tried to generalize the results concerning the problem of loop condensation in disorder array of topological obstacles (considered in Appendix Section 1.1) to the case of complex loops. We have showed that the general result of I remains unchanged, i.e. random distribution of obstacles induces the collapse transition in the loop prepared in specific topological state. However the phase behavior becomes much more intricate. We have found, for instance, that the collapse transition is accompanied by the rebuilding of internal structure of the loops from symmetric to asymmetric distribution in the leaf space (see above).

We believe that the proposed model can be also used for the investigation of the stress–strain behaviour of polymer chains in random arrays of topological obstacles what could be regarded as the basis for the description of high elastic properties of irregular gels, which demonstrate very unusual thermodynamic behavior under deformations [24,25].

References

1. S.K. Nechaev, V.G. Rostiashvili, J. Phys. II (France), 3 (1993), 91
2. V.G. Rostiashvili, S.K. Nechaev, T.A. Vilgis, Phys. Rev.(E), 48 (1993), 3314
3. F.A.L. Dullein, *Porous Media, Fluid Transport and Pore structure*, (Academic Press: N.Y., 1979)
4. W.W. Yan, J.J. Kikland, D.E. Bly, *Modern Size Exclusion Liquid Cromatography*, (J.Wiley: N.Y., 1979)
5. A. Baumgarthner, M. Muthukumar, J. Chem. Phys., 87 (1987), 3082
6. S.F. Edwards, M. Muthukumar, J. Chem. Phys., 89 (1988), 2435
7. M.E. Cates, R.C. Ball, J. Phys. (France), 49 (1988), 2009
8. T.A. Vilgis, J. Phys. (France), 50 (1989), 3243
9. A. Baumgarthner, M. Moon, Europhys. Lett., 9 (1989), 203
10. F.F. Ternovskii, A.R. Khokhlov, Zh.Exp.Teor.Fiz., 90 (1986), 1249
11. A.R. Khokhlov, F.F. Ternovskii, E.A. Zheligovskaya, Physica A, 163 (1990), 747
12. S.V. Panykov, Zh.Exp.Teor.Fiz., 94 (1988), 174
13. P. Goldbard, N. Goldenfeld, Phys. Rev. (A), 39 (1989), 1402, 1412
14. M.G. Brereton, S. Shah, J.Phys. (A), 13 (1980), 2751; D.J. Elderfield, J. Phys. (A), 15 (1982), 1369; F. Tanaka, Progr. Theor. Phys., 68 (1982), 148, 164.
15. V.J. Emery, Phys.Rev.(B), 11 (1975), 239
16. S. Coleman, E. Weinberg, Phys. Rev. (D), 7 (1973), 1888

17. W. Horsthemke, R. Lefever, *Noise-Induced Transitions*, (Springer: Berlin, 1984)
18. S. Nechaev, Int. J. Mod. Phys. (B), 4 (1990), 1809
19. A.R. Khokhlov, S.K. Nechaev, Phys. Lett. (A), 112 (1985), 156
20. R.T. Deam, S.F. Edwards, Philos. Trans. R. Soc. London (A), 280 (1976), 317
21. G. Ferry, *Viscoelastic Properties of Polymers* (Wiley: New York, 1980)
22. S.F. Edwards, in *Polymer Networks*, edited by A.J. Chompff and S. Newman (Plenum Press: New York, 1972)
23. R.P. Feynman and A.R. Hibbs, *Path Integrals and Quantum Mechanics* (Academic Press: New York, 1965)
24. Y. Rabin, R. Bruinsma, Europhys. Lett., 20 (1992), 79
25. S.V. Panyukov, ZhETP Letters, 58 (1993), 114 (in Russian)

Subject Index